# MECHANICAL PROPERTIES
# OF REINFORCED THERMOPLASTICS

# MECHANICAL PROPERTIES OF REINFORCED THERMOPLASTICS

*Edited by*

## D. W. CLEGG

*Department of Metals and Materials Engineering, Sheffield City Polytechnic, UK*

and

## A. A. COLLYER

*Department of Applied Physics, Sheffield City Polytechnic, UK*

ELSEVIER APPLIED SCIENCE PUBLISHERS
LONDON and NEW YORK

ELSEVIER APPLIED SCIENCE PUBLISHERS LTD
Crown House, Linton Road, Barking, Essex IG11 8JU, England

*Sole Distributor in the USA and Canada*
ELSEVIER SCIENCE PUBLISHING CO., INC.
52 Vanderbilt Avenue, New York, NY 10017, USA

WITH 56 TABLES AND 99 ILLUSTRATIONS

© ELSEVIER APPLIED SCIENCE PUBLISHERS LTD 1986

**British Library Cataloguing in Publication Data**

Mechanical properties of reinforced thermoplastics.
1. Thermoplastics
I. Clegg, D. W.     II. Collyer, A. A.
668.4′23     TP1180.T5


**Library of Congress Cataloging-in-Publication Data**

Mechanical properties of reinforced thermoplastics.

Bibliography: p.
Includes index.
1. Reinforced thermoplastics — Mechanical properties.
I. Clegg, D. W.   II. Collyer, A. A.
TA455.P55M43   1986     620.1′923          85-31162

ISBN 0-85334-433-7

The selection and presentation of material and the opinions expressed in this publication
are the sole responsibility of the authors concerned.

### Special regulations for readers in the USA

Phototypesetting by Tech-Set, Gateshead, Tyne & Wear.
Printed in Great Britain by Galliard (Printers) Ltd, Great Yarmouth.

# Preface

The reinforcement of materials such as mud and clay by hair, straw and vegetable fibres has been long established in man's history, enabling him to improve his buildings and extend his engineering abilities. With the advent of modern synthetic polymers it was rapidly realised that the addition of fibres, flakes and particulate materials to polymer matrices could improve mechanical properties significantly. Fibres and flakes are the most effective and have enabled several polymers with limited properties to compete with long-established metallic materials, resulting in cost, weight and processing economies. This is increasingly apparent in the selection of materials for aerospace and road vehicle applications as well as in a multitude of domestic products.

Reinforced plastics, both thermosets and thermoplastics, are used in increasingly harsh environments involving elevated temperatures and aggressive conditions. Fibre reinforcement of thermoplastics dominates, and a pattern of increasing replacement of fibre reinforced thermosets by reinforced thermoplastics is emerging. This trend is encouraged by the development of continuous fibre reinforced grades of the newer high-temperature engineering thermoplastics such as polyether ether ketone.

The first part of this book reviews the mechanical properties and theories of short fibre reinforcement. The principal reinforcements are reviewed and a separate chapter is devoted to the uses of natural fibres as reinforcements for thermoplastics. This is an interesting and commercially important area, especially for Third World countries where these fibres are grown but are facing severe competition from synthetic fibres in traditional applications such as ropes and matting. Subsequent chapters cover the exciting developments in continuous fibre reinforcement of thermoplastics and the requirements, theories and technologies involved in the bonding between reinforcements and matrices.

The second part of the book concerns the interrelationship between constitution, morphology, rheology and processing characteristics of reinforced thermoplastics. It covers the most recent developments and discusses techniques used to study the materials.

In addition, a chapter is devoted to designing with short fibre reinforced thermoplastics and the influence of fibre orientation on design is explained.

The work as a whole shows the importance of many varied disciplines to the understanding and development of reinforced thermoplastics, and will be of interest to materials scientists, engineers and technologists in industry, research laboratories and academic institutions. It is a worthwhile introduction to the subject for students as well as providing an up-to-date review of recent innovations in the field.

Finally, we would like to thank Mr J. Evans for drawing many of the diagrams, Mr M. Morris for supplying Fig.1.2 and Dr A. Norcliffe for helpful discussions on mathematical theories. Other acknowledgements are given at appropriate places in the text.

D. W. CLEGG and A. A. COLLYER

# Contents

# List of Contributors

R. H. BURTON
*Department of Materials Technology, Brunel University, Kingston Lane, Uxbridge, Middlesex UB8 3PH, UK*

D. W. CLEGG
*Department of Metals and Materials Engineering, Sheffield City Polytechnic, Pond Street, Sheffield S1 1WB, UK*

F. N. COGSWELL
*Imperial Chemical Industries PLC, New Science Group, PO Box 90, Wilton, Middlesbrough, Cleveland TS6 8JE, UK*

A. A. COLLYER
*Department of Applied Physics, Sheffield City Polytechnic, Pond Street, Sheffield S1 1WB, UK*

C. A. CRUZ-RAMOS
*Chemistry Division, Polymers Department, Centro de Investigacion Científica de Yucatán, AC, Apdo. Postal No 87 Cordemex, Yucatán 97310, Mexico*

M. W. DARLINGTON
*Department of Materials, Cranfield Institute of Technology, Cranfield, Bedford MK43 0AL, UK*

ix

M. J. FOLKES

*Department of Materials Technology, Brunel University, Kingston Lane, Uxbridge, Middlesex UB8 3PH, UK*

L. A. GOETTLER

*Monsanto Chemical Company, 260 Springside Drive, Akron, Ohio 44313, USA*

G. C. MCGRATH

*Department of Metals and Materials Engineering, Sheffield City Polytechnic, Pond Street, Sheffield S1 1WB, UK*

E. P. PLUEDDEMANN

*Dow Corning Corporation, Midland, Michigan 48640, USA*

H. W. RAYSON

*Department of Metals and Materials Engineering, Sheffield City Polytechnic, Pond Street, Sheffield S1 1WB, UK*

P. H. UPPERTON

*Engineering Plastics Materials Division, Du Pont (UK) Ltd, Maylands Avenue, Hemel Hempstead, Herts HP2 7DP, UK*

J. L. WHITE

*Center for Polymer Engineering, College of Engineering, University of Akron, Akron, Ohio 44325, USA*

*Chapter 1*

# An Introduction to Fibre Reinforced Thermoplastics

A. A. COLLYER and D. W. CLEGG

*Polymer Group, Sheffield City Polytechnic, Sheffield, UK*

## 1.1. INTRODUCTION

When considering materials for load-bearing applications, designers are increasingly examining the advantages of using plastics materials, both thermosets and thermoplastics, instead of traditionally accepted materials and, in particular, metals. Thermoplastics have three main advantages, their low specific gravities, their low energy requirements for manufacture and their low costs of fabrication, particularly by the injection moulding route.

The strength-to-weight and stiffness-to-weight ratios of various materials are shown in Table 1.1.[1,2] The thermoplastics have higher strength-to-weight ratios than aluminium and steel, but lower stiffness-to-weight ratios, giving rise to increased buckling under load. Therefore, when designing with thermoplastics, the design is influenced more by stiffness than by strength.

When energy costs are high, the amount of energy required to produce unit volume of material is an important consideration. It can be seen from Table 1.2[3] that the energy required to produce a cubic metre of thermoplastics is less in general than that required for metal production. Environmentalists will approve of this.

The third point of comparison, namely fabrication, sees another advantage offered by thermoplastics materials. They can be made rapidly and reproducibly into complex shapes by injection moulding,

TABLE 1.1
Comparison of the Mechanical Properties of Several Types of Engineering Materials[1,2]

| Material | Specific gravity $(g\,cm^{-3})$ | Modulus $(GN\,m^{-2})$ | Specific modulus $(MN\,m\,kg^{-1})$ | Strength $(MN\,m^{-2})$ | Specific strength $(kN\,m^{-1}\,kg^{-1})$ |
|---|---|---|---|---|---|
| Aluminium | 2·7 | 71 | 26 | 80 | 30 |
| Brass (70Cu/30Zn) | 8·5 | 100 | 12 | 550 | 65 |
| Mild steel | 7·86 | 210 | 27 | 460 | 59 |
| Polyamide 66 | 1·14 | 3 | 2·6 | 80 | 70 |
| Polycarbonate | 1·24 | 2·3 | 1·9 | 60 | 48 |
| Polyamide 66/30% glass | 1·38 | 8 | 5·8 | 160 | 116 |

TABLE 1.2
Energy Requirements for the Production of Different Materials
(After Sheldon[3])

| Material | Energy | |
|---|---|---|
| | $(MJ\,kg^{-1})$ | $(KJ\,m^3/10)$ |
| Bottle glass | 18 | 41 |
| Low-density polyethylene | 69 | 64 |
| High-density polyethylene | 70 | 67 |
| Polypropylene | 73 | 68 |
| Polyvinyl chloride | 53 | 69 |
| Polystyrene | 80 | 84 |
| Polyurethane | 130 | 100 |
| Polypropylene/30% glass fibre | 90 | 100 |
| Polyester/30% glass fibre | 90 | 150 |
| Phenoplast | 150 | 200 |
| Steel | 45 | 350 |
| Aluminium | >200 | >540 |
| Brass | 95 | 600 |

which is not labour intensive and lends itself well to microprocessor control.

The importance of thermoplastics has increased since the advent of the newer, high-temperature engineering thermoplastics; various polyamides, polysulphone (Union Carbide Ltd), polyether sulphone (ICI), polyether ether ketone (ICI), polyether imide (General Electric) and polyimide (DuPont). All but the last can be injection moulded. Table 1.3[4–9] shows typical properties of some of these materials. These mechanical properties compare well with those of aluminium, even at elevated temperatures (200°C), resulting in a number of the materials being used where traditionally a metal would have been chosen. Often this has led to an improvement in design of the part.

One example of this is in the use of polyether sulphone for printed circuit boards. This material can be soldered with no distortion, and the production of the parts by injection moulding is rapid and cheap. The new improvement offered by the use of this material is that of three-dimensional printed circuit boards.[10]

TABLE 1.3
Properties of Various High-temperature Engineering Thermoplastics Compared with Aluminium[4-9]

| Property | Unit | Polysulphone | Polyether sulphone | Polyether ether ketone | Polyether imide | Polyimide | Aluminium |
|---|---|---|---|---|---|---|---|
| Specific gravity | $g\,cm^{-3}$ | 1·24 | 1·37 | 1·32 | 1·27 | 1·36 | 2·7 |
| Tensile modulus | $GN\,m^{-2}$ | 2·5 | 2·44 | 1·1 | 3·0 | 1·43 | 71 |
| Tensile strength at yield (20°C) | $MN\,m^{-2}$ | 70 | 80 | 91 | 105 | 72–86 | 80 |
| Flexural strength (20°C) | $MN\,m^{-2}$ | 106 | 129 | — | 200 | — | — |
| Notched Izod impact | $J\,m^{-1}$ | 69 | 84 | — | 50 | 80 | — |

| Property | Units | | | | | | |
|---|---|---|---|---|---|---|---|
| Charpy notched impact | kJ m⁻² | — | — | 10·0 | 54 | — | — |
| Rockwell hardness | Class | M69 | M88 | M109 | — | — | B70 |
| Glass transition temperature | °C | 190 | 225 | 230 | 143 | — | — |
| Melting point | °C | — | — | — | 334 | — | 660 |
| Heat deflection temperature (1·85 MN m⁻²) | °C | 173 | 203 | 200 | 135–160 | 360 | — |
| Vicat softening point (1 kg) | °C | — | 226 | 219 | — | — | — |
| Upper service temperature | °C | 171 | 180 | 170 | 260 | 315 | 100 |

If thermoplastics are so good, why have they not completely replaced metals? One of the main reasons is their poor high-temperature capability. If there were three main improvements to make to thermoplastics, they would be to stiffness, strength and creep resistance at elevated temperatures. It is instructive to consider how modifications can be made to achieve this end.

## 1.2. METHODS OF IMPROVING THE MECHANICAL PROPERTIES OF THERMOPLASTICS

There are several methods of achieving improved mechanical properties; in describing them it is hoped to place in context the role of fibre reinforced thermoplastics.

### 1.2.1. Molecular Architecture

Manipulation of the molecular architecture of a polymer necessarily means developing a new material and the industrial process by which it can be manufactured in bulk. This is extremely speculative and costly, and new thermoplastics, unless markedly different from existing ones, have a limited potential for penetrating the market.

This has not always been true and it is useful to examine how the different architectures developed over the years influence stiffness and give higher temperature capabilities. Figure 1.1(a) shows the molecular configuration of high-density polyethylene (HDPE). It has poor mechanical properties and cannot be used at high temperatures; as such it is unsuitable for engineering applications. The molecule has a flexible backbone chain that can extend and rotate into many conformations with relative ease, giving a low melting temperature, $T_m$, and a low stiffness.

A better material from a load-bearing viewpoint, polystyrene, is shown in Fig. 1.1(b). Here, the aromatic sidegroups restrict the movement of the backbone chain, giving a far more rigid material. As it happens, polystyrene is very brittle and not suited to engineering applications in unmodified form.

The secret of a higher-temperature capability and stiffness lies partly in the ease of movement of the backbone chain. In polyphenylene oxide (PPO), Fig. 1.1(c), the aromatic rings are in the backbone structure, giving a greatly restricted movement. Unfortunately, the restriction is too much, and PPO is an intractable material, although its mechanical

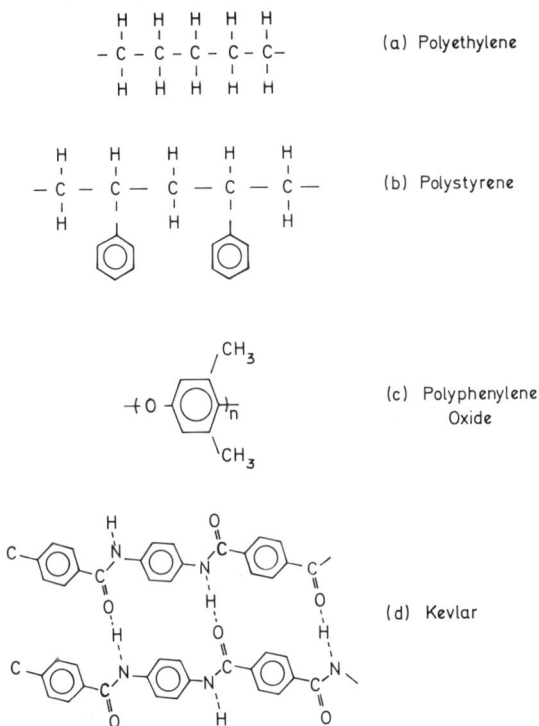

**Fig. 1.1.** Molecular architecture of some thermoplastics.

properties are desirable. PPO is usually modified by incorporating a graft copolymerisation with PS. This is available as 'Noryl' (General Electric Plastics).[11]

Figure 1.1(d) shows the aromatic polyamide structure of Kevlar 49, another intractable material. This is used as a reinforcing fibre in thermoplastics matrices, and is included for that reason. Often, intractable materials decompose before they melt.

In this section, so far, the high-temperature and stiffness characteristics have been related only to the overall structure; these desirable characteristics can be achieved by replacing the carbon–carbon bonds by stronger ones, such as silicon–oxygen. Eventually this would lead to a discussion of ceramic materials, many of which are used as reinforcements. Further discussion of molecular architecture is given by Hearle[12] and Mascia.[13]

### 1.2.2. Copolymerisation

Owing to the expense in the design and manufacture of new polymers, research has been carried out to find synergistic properties in blends and copolymers of existing materials. Blends, in general, are not very useful because the incompatibility between the constituents gives rise to poor adhesion at the interface and poor mechanical properties. In high shear rate processes, such as injection moulding, phase separation occurs as shown in Fig. 1.2.

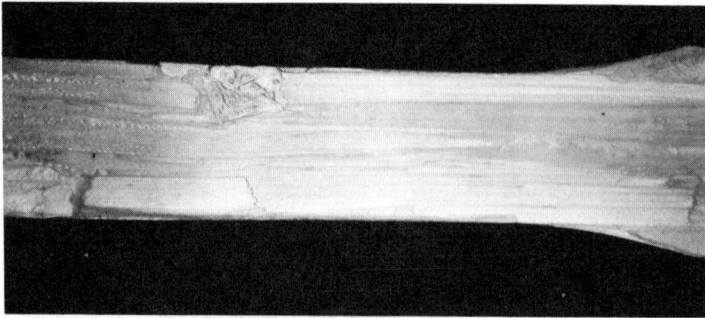

Fig. 1.2. Delamination in an injection moulding from a blend of polyether-sulphone and polydimethylsiloxane. (Courtesy of M. Morris, Sheffield City Polytechnic.)

A much better material is obtained if the mixing is accomplished at a chemical rather than a mechanical level. This will give a much better adhesion between the constituents, and is achieved by using a random, a graft or a block copolymer. The latter two are particularly good as compatibilisers at the interfaces between the two constituent homopolymers. The copolymerisation is carried out with both the homopolymers, species A and species B, present, and structures such as those shown in Fig. 1.3 are obtained. The behaviour of the copolymer depends on its type and on the nature and concentration of its constituents. One of the most common examples of a copolymer is ABS, which is available in many grades. 'Noryl' has already been mentioned.

Propylene and ethylene are often copolymerised in almost any proportions to give a spectrum of grades varying from rigid thermoplastics to thermoplastic elastomers. The main difference between the copolymer and the two homopolymers is that the first has a better low-temperature shock resistance at the cost of a slight loss in rigidity and hardness at higher temperatures.

A — A — B — A — B — B — A — B — B — A — A    (a) Random

B
 ╱
B
 ╱
A — A — A — A — A — A — A — A    (b) Graft
        ╲
         B
          ╲
           B
            ╲
             B

A — A — A — A — A— A — B - B - B - B - B - B - B
                                             (c) Block

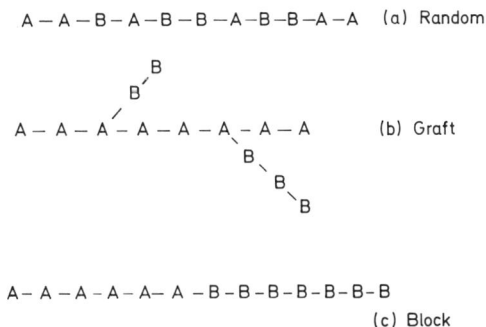

**Fig. 1.3.** Random, graft and block copolymers.

Work carried out by Noshay *et al.* [14, 15] has shown that polysulphone (PSU) can be toughened by the incorporation of a block copolymer of PSU with polydimethylsiloxane (PDMS) in the PSU matrix. Again, the copolymer approach necessitates the design of a process that can produce the material on an industrial scale.

### 1.2.3. Crystallinity

Another way of improving stiffness and promoting good high-temperature mechanical properties is to encourage crystallinity. The alignment of chains gives regions of crystallinity, which reduce the chain movement. The melting point is high when the intermolecular forces are high. This can be achieved by increasing the polarity of constituent groups or elements and reaches a maximum if hydrogen bonding can be achieved between hydrogen atoms in one chain and unpaired electrons from an electronegative atom in another chain. The different forms of polyethylene show varying degrees of crystallinity. The inclusion of sidechains reduces the chance of crystallinity; the most crystalline form of polyethylene is, therefore, the linear HDPE, which can be used to higher temperatures than LDPE. Reducing chain branching to promote crystallinity again requires a manipulation of the molecular architecture. Crystallinity is also reduced by chain stiffness and a lack of symmetry; the former reduces the motion necessary for alignment of chains, and the latter reduces the ability to line up.

### 1.2.4. Crosslinking

This method represents the first of the cheaper methods of improving stiffness. A crosslinked material such as a vulcanised rubber or a thermoset is one which derives its stiffness from the fixed nature of the crosslinks. Such a material will not flow once it has undergone the crosslinking action and will lose its dimensional stability when it begins to char. Only recently have thermoplastics rivalled thermosets in high-temperature capability, and the thermoplastics materials have the advantages of faster, easier processing and the possibility of being recycled.

Thermosets have been successfully and widely used as matrices for reinforcement: many recent texts have been written concerning them.[16-20]

## 1.2.5. Reinforcement

The reinforcement of thermoplastics and thermosets by ceramic, metallic or polymeric fibres is of vast importance. It is easily achieved and often the composite formed is cheaper than the polymer matrix alone, an unexpected advantage perhaps. These reinforced grades marry the strength and stiffness of such fibres with the good shock resistance of the thermoplastic matrix. The fibres alone are usually very brittle and their strength and stiffness cannot be fully realised. The matrix protects these fibres and transfers the load to them. This gives a material that combines the good properties of the fibre and the matrix, producing an improvement in the strength, stiffness and creep resistance over those properties for the matrix alone. These composite materials offer good competition to metals in many applications in the motor and aerospace industries and in domestic appliances.

The strengthening mechanism depends on the geometry of the reinforcing filler, which may be one of two types, particulate or fibrous. A particulate filler has no long dimension, platelets being a noted exception. As a long dimension discourages the growth of incipient cracks normal to the reinforcement in a brittle matrix, a particle does not improve the fracture toughness of such a matrix. The exception to this rule is when a rubberlike substance is dispersed in a brittle matrix. Under these conditions, considerable toughening occurs, and this method is standard for improving the impact behaviour of thermoplastics. Typical examples are high-impact polystyrene and ABS.

The particles will also share the load with the matrix, but to a lesser extent than a fibre. A particulate reinforcer will, therefore, improve stiffness but will not generally strengthen. Hard particles in a brittle matrix will cause localised stress concentrations in the matrix, which will reduce the overall impact strength.

— Particulate fillers are employed to improve high-temperature performance, reduce friction, increase wear resistance, improve machinability and reduce shrinkage. In many cases, particulate fillers are used simply to reduce cost, and as such their study is outside the scope of the present work. Under these conditions the additive is a filler, whereas when a considerable change in properties of the composite occurs, the additive is a reinforcement.

Fibre reinforcement improves the three chief weaknesses of a thermoplastic matrix, stiffness, strength and creep resistance. The measured strength of most materials is much less than that predicted by theory because flaws in the form of cracks perpendicular to the applied load are present in bulk materials. Fibres of non-polymeric materials have much higher longitudinal strengths in this form because the large flaws are not generally present in such small cross-sectional areas. However, these small cross-sectional areas do not permit the use of fibres alone in engineering applications. In the case of polymeric materials, such as Kevlar, the orientation of the polymer molecules along the long dimension produces strength in that direction. Table 1.4[19] shows the mechanical properties of fibres used in reinforcement. E-glass is the most commonplace because of its relatively low cost. Boron, carbon and polyaramid fibres (Kevlar 49) are important because of their high stiffness. Carbon fibres offer the greatest versatility owing to their great variety and the ease of control of their structure.

TABLE 1.4
Typical Properties of Fibres Used for the Reinforcement of Thermoplastics[19]

| Fibre | Density $(g\,cm^{-3})$ | Young's modulus (average value) $(GN\,m^{-2})$ | Tensile strength (average value) $(GN\,m^{-2})$ |
|---|---|---|---|
| E-glass | 2·55 | 75 | 2·0 |
| S-glass | 2·49 | 75 | 5·5 |
| Carbon | 2·00 | 170–200 | 0·5–1·0 |
| Boron | 2·60 | 400–450 | 3·0–3·5 |
| Chrysotile asbestos | 2·50 | 160 | 2·0 |
| Silicon carbide | 3·15 | 220–300 | 4·0–10·0 |
| Kevlar | 1·45 | 130 | 3·0–3·6 |

The fibres dispersed in the matrix may be continuous, in which case it is easy to imagine how the load is transferred to them by the matrix, or they may be short. In the latter case, the fibres must be of sufficient length to have the load transferred to them efficiently. The fibre lengths are in the range 0·125–0·5 mm. These short fibre reinforced thermoplastics have become of paramount importance in replacing metals in applications requiring a combination of strength and lightness.

Table 1.5[21] shows the improvement of fibre reinforcement over the matrix material. Clearly a division is observable here; the fibre reinforcement greatly enhances the continuous service temperature of crystalline polymers but only slightly increases it in amorphous ones.

TABLE 1.5
Benefits of Reinforcement with 30% W/W Glass Fibres of Crystalline and Amorphous Plastics[21]

| Plastics material | Heat distortion temperature at 1·81 MN M$^{-2}$ (°C) | | Tensile strength at 23°C (MN m$^{-2}$) | |
|---|---|---|---|---|
| | Actual | Enhancement | Actual | Enhancement |
| *Crystalline* Nylon 66 | 248 | +153 | 180 | +100 |
| Polyether ether ketone | 300 | +145 | 175 | +75 |
| Nylon 6 | 212 | +137 | 160 | +100 |
| Polypropylene | 148 | +83 | 86 | +51 |
| *Amorphous* Polyether sulphone | 216 | +15 | 145 | +55 |
| Noryl | 145 | +15 | 125 | +59 |
| Polycarbonate | 140 | +10 | 120 | +56 |
| ABS | 100 | +10 | 90 | +40 |

Fibre reinforcement represents a physical rather than chemical means of changing a material to suit various engineering applications; this is far cheaper than redesigning thermoplastics by a chemical route.

## 1.2.6. Foaming

Stiffness of a structure can be achieved by producing a foam, such that a thicker section is used of the same mass as the original. This method, however, lowers the continuous working temperature of the composite. There are several kinds of foamed composite:

(1) a composite comprising a gas dispersed in the polymer matrix;
(2) a macrocomposite similar to (1) in which the walls are thick and the microcellular structure is confined to the centre of the moulding;
(3) syntactic foams which consist of a dispersion of rigid hollow microspheres in a polymer matrix; and
(4) reinforced foams using fibre reinforcement of the matrix containing the microcellular structure.

Microcellular foams can be produced on normal injection moulding machines by incorporating a blowing agent into the feedstock. On being heated, the blowing agent forms the dispersed gas phase. Density reductions of up to 30% can be achieved in this way. Specialised injection moulding machines produce better results, and in the case of polyurethane foams a new process, reaction injection moulding (RIM), was developed and has been successfully used with other materials.

Thermoplastic materials that may be foamed include PS, ABS, polypropylene, polyethylene, nylon, PVC and polycarbonate. This list is by no means exhaustive.

The thickness of the walls of the foam may be controlled and may even be of a different polymer from that used in the microcellular foamed centre, provided the two polymers adhere well to each other. Such a composite is sandwich moulded in one operation in an injection moulding machine.

Applications of the microcellular foams and the sandwich moulded foams include fan shrouds, furniture components, garden frames, television cabinets, motor vehicle fascias and bumpers and building panels.

Syntactic foams consist of rigid, hollow, inorganic microspheres, such as glass balloons of fly ash, dispersed in a polymer matrix. Sometimes organic microspheres made from phenol, urea–formaldehyde, polyvinylidene chloride or expandable polystyrene are used. The last are fabricated as microspheres containing a blowing agent that can expand at the softening point of polystyrene. The diameters of the microspheres used in syntactic foams are in the range 30–120 $\mu$m, with wall thicknesses between 2 and 3 $\mu$m. These foams are stronger than their unfilled counterparts and can withstand higher hydrostatic pressures. They are used as buoyancy aids in ships and submarines, as buoys, cones and rudders, and in camping equipment, in electronic equipment because of their low electrical loss properties and in simulated lightweight wood applications. The radome on the MIG fighter plane is made from hollow glass microspheres in a polyether sulphone matrix.

The reinforced foams are of particular interest. They can be produced in a number of ways: by adding expandable polystyrene beads to a reinforced polymer, by introducing nitrogen gas into a reinforced polymer, or by using a blowing agent. One specialised process called reinforced reaction injection moulding (RRIM) was developed from RIM to produce short fibre reinforced microcellular foams. Typical uses of these foams are in fruit cases, car seating, automobile panels and in self-sealing fuel tanks in military vehicles.

The whole subject of foams is vast and is described by Hilyard.[22]

## 1.3. COUPLING AGENTS

The success of glass fibre reinforced polyamides is due partly to the good adhesion between the two species. Olefinic materials do not bond well with glass, and other thermoplastic matrices need coupling agents — otherwise the fibres tend to slip out of engagement under load and benefits of their presence are lost.

Inorganic fillers usually contain hydroxyl groups on their surfaces owing to reactions with atmospheric water or to the existence of strong adsorptive forces caused by high surface energy. Only those thermoplastics that can form hydrogen bonds are capable of bonding strongly to these materials.

Moreover, the hydrophilic nature of the inorganic fillers attracts atmospheric water to the interface with the matrix, resulting in a deterioration of mechanical properties with time in humid environments. Coupling agents are, therefore, crucial to the strengthening mechanism in fibre reinforced thermoplastics.

The most common coupling agents are 'silane'- and 'titanate'- based compounds, whose chemical composition allows them to react both with the surface of the fibre and with the polymer matrix. The make-up of the organic part of the molecules of a coupling agent can be tailored to suit the polymer matrix to be used. For silane coatings, about 1% of coupling agent is used which will give a coating of thickness 0·05 μm on a fibre of 10 μm diameter. The coatings are applied to the fibres before they are compounded with the polymer matrix.

The titanate coupling agents are exemplified by isopropyl tri(dioctyl pyrophosphate) titanate, which binds to an inorganic surface by hydrolysis.[23] Figure 1.4 shows a scheme by which the titanate coupling agent reacts with the surface of the fibre,[13] and Table 1.6 shows three typical titanate coupling agents, their composition and applications.

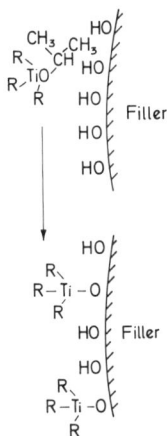

Fig. 1.4. Reaction of titanate coupling agents with the surface of the reinforcing fibre.[13]

Among the non-silane coupling agents are the chromium complexes supplied as 'Volan' by DuPont. They have been used extensively with glass: the hydroxyl groups bind to the silanol groups of the glass surface through hydrogen bonding and possibly covalent oxane bonding.[24] The organic acid group develops a fairly stable bond with chromium by being coordinated to adjacent chromium atoms. These coupling agents have been used in glass reinforced polyesters and epoxies. Other specialised non-silane coatings include phosphorus-containing compounds, long-chain aliphatic carboxylic acids and certain amines, but their use is not widespread.

Polymers have also been used as coupling agents between two other homopolymers. An example of this is the use of polyurethane to couple aramid fibres to an epoxy matrix. Many of the polymer coupling agents have been used with a view to improve the shear and impact behaviour

TABLE 1.6
Typical Titanate Coupling Agents[13]

| Chemical name | Chemical structure | Applicable matrices |
|---|---|---|
| Monoalkoxy isostearoyl dimethacrylate titanate | | Polyolefins |
| Monoalkoxy tri(dioctyl pyrophosphate) titanate | | Vinyl polymers |
| Monoalkoxy 4-aminobenzenesulphonyl di(dodecylbenzenesulphonyl) titanate | | Polyamides |

of the matrix. Some examples of these systems are silicone-rubber coated carbon fibre, epoxy coated boron, and polyester coated glass for polyester matrices. Particulate fillers impregnated with polyethylene have been much used. In these systems the thickness of the coating is of great importance, the optimum thickness being of the order of microns, which is much thicker than the conventional silane coatings.

A different type of coupling is exemplified by graft or block copolymers. These are used as compatibilising agents at the interface of two immiscible homopolymers. Their use is not generally associated with fibre reinforced thermoplastics.

The study of coupling agents is a science in itself and is covered by a number of texts[2, 13, 24, 25] as well as in Chapter 8.

## 1.4. FIBRE REINFORCED THERMOPLASTICS

Originally the purpose of fibre reinforcement was to improve the strength, stiffness and creep resistance of polymer matrices. It is interesting to analyse how well these intentions have been realised, to consider any additional, perhaps unexpected, advantages accruing due to reinforcement, and to study the disadvantages that are involved in reinforcement.

### 1.4.1. Strength and Flexural Modulus
Plastics are not as strong as metals on a strength per unit volume basis, but the incorporation of fibres into polymer matrices increases the strength and the heat deflection temperature (Table 1.7).

The flexural modulus of plastics materials is an order of magnitude lower than that of traditional construction materials, but the incorporation of suitable fibres and coupling agents considerably reduces the differences between the two types of material.

### 1.4.2. Creep Resistance and Temperature Resistance
Figure 1.5 shows the effect of fibre reinforcement on creep. Clearly, the reinforcement reduces creep and enables the composite to be used at higher temperatures until creep becomes unacceptable.

The upper limit for continuous operation is set by polyimide at 315 °C and by polyether ether ketone at 260 °C. The former is intractable and cannot be injection moulded. As mentioned earlier, the reinforcement is far more effective in increasing the upper service temperature in crystalline materials (Table 1.5).

TABLE 1.7
Properties of Commercial Grades of Polyether Ether Ketone[6]

| Property | PEEK | PEEK +20% w/w glass | PEEK +20% w/w carbon |
|---|---|---|---|
| Specific gravity (g cm$^{-3}$) | 1·3 | 1·37 | 1·4 |
| Tensile modulus at 150°C (GN m$^{-2}$) | 1·1 | | |
| Tensile strength at yield at 25°C (MN m$^{-2}$) | 100 | 149 | 165 |
| Flexural strength (MN m$^{-2}$) | | | 260 |
| Notched Izod impact strength (kJ m$^{-2}$) | 54 | 7·6 | 0·048 |
| Heat deflection temperature at 1·81 MN m$^{-2}$ (°C) | 135–160 | 286 | 300 |
| Continuous service temperature (°C) | 250 | — | — |

The reinforcement with a suitable coupling agent will therefore achieve the three main objectives of improving the mechanical properties of thermoplastics, but there are other benefits too.

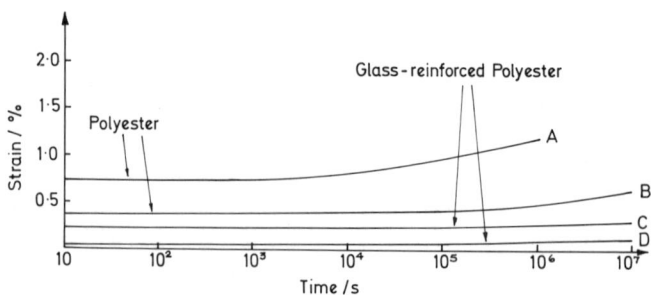

Fig. 1.5. Creep behaviour of reinforced and unreinforced grades of polyesters at 20°C. Tensile stress: for the unreinforced grades, (A) 20 MN m$^{-2}$, (B) 10 MN m$^{-2}$; for reinforced grades, (C) 20 MN m$^{-2}$, (D) 10 MN m$^{-2}$.

### 1.4.3. Surface Hardness

Thermoplastics have inferior surface hardness to metals and glass. Poly(methyl methacrylate) (PMMA) displays the best surface hardness, which is similar in performance to that of aluminium. The reinforcement of thermoplastics actually reduces surface hardness. Schmitt[26] examined the erosion of the surfaces of plastics under the action of supersonic raindrops and found that the erosion rate was increased by reinforcing fibres, but a similar reinforcement in thermosets reduced the erosion of the surface. Practically all the ultra-high-temperature resins such as polyimide, polybenzimidazole and silicone are prone to surface erosion. Performances are improved by reducing voids and by using glass fibres rather than boron or carbon.[28]

There are no modifiers for improving surface hardness, except for the incorporation of graphite into nylon, which increases surface hardness by increasing the crystallinity in the surface layer of the polymer.[28]

A coating of silicone materials will improve surface hardness, as will elastomeric polyurethane coatings. The latter will resist high-velocity raindrop erosion, but the impact stresses are transmitted to the substrate, which erodes as if unprotected.

Exposure to high-velocity raindrops occurs in aerospace applications such as radomes and high-speed turbine blades.

### 1.4.4. Resistance and Outdoor Weathering

The acrylics and the new engineering thermoplastic, polyetherimide (PEI), show excellent resistance to UV light and weathering in tropical conditions. However, polystyrene and polypropylene without modification deteriorate rapidly to low strengths after a few months' exposure to strong sunlight. The addition of fibres makes little difference here, although carbon fibres should be preferable to glass,[29] but fortunately the matrices can be greatly improved by the addition of UV stabilisers.

The important wavelengths lie in the range 280–400 nm; this radiation breaks many chemical bonds giving rise to the formation of free radicals, which cause a complex series of degradation reactions. The UV action is purely a surface one, but a short period of exposure to high climatic temperatures can extend the depth of the reaction zone.

### 1.4.5. Flammability

All plastics materials that are polymers of hydrocarbons will burn. The chemical degradation leads to the production of inflammable gases, and often a great deal of dark smoke is generated, especially from highly carbonaceous polymers such as those containing substantial aromatic matter. A review of this area of study is given by Wilson.[29]

Grades with flame retardant additives are readily available. Care must be taken with these grades, as often the additive will merely increase the temperature at which ignition occurs, and thereafter will add to the toxicity of the fumes and smoke produced. Recent examples show that it is preferable to use a more fire resistant polymer than PVC, such as PEEK, as a wire-covering material in applications where fire is a real hazard. Recent experiments with a blowtorch on aluminium and a PEEK/carbon fibre composite showed that the latter withstood the flame much better.[10]

The addition of fibres often improves the fire resistance marginally, but glass fibres in polyamides reduce their fire resistance because the increase in the viscosity of the composite reduces the heat transfer by dripping. Glass microspheres do not show this effect to the same extent.[27]

An area of thermal breakdown of particular interest is that of ablation, which has come into prominence with regard to re-entry problems in space flights. Temperatures of 5000–15 000 °C may be generated for short periods of time on re-entry. Thermoset composites have been used in a sacrificial role to char progressively: in so doing, they release cooling gases that delay the progress of the heat pulse into the centre of the spacecraft. The fibre reinforcement chosen often breaks down producing additional cooling gases to add further heat resistance.[3]

### 1.4.6. Thermal Expansion and Mould Shrinkage

The thermal expansion of plastics varies between five and ten times that of metals. This can cause problems when the material cools in an injection mould. Allowances is made for the shrinkage in the design stage. Crystalline materials show mould shrinkages of the order of 1–4% and the shrinkage may not be isotropic. Amorphous materials shrink less than this, 0·5–1·0%, with a uniform shrinkage in all directions.

Fibre reinforcement reduces mould shrinkage, but when there is a high degree of orientation the shrinkage will be less along the fibre axes. Bead filled grades show uniform shrinkage and are less likely to warp.

### 1.4.7. Electrostatic Charges

Polymers are poor conductors of electricity and are prone to a build-up of static charges on their surfaces. This is particularly noticeable on some artificial fibred carpets. At times it can be dangerous, for example when a build-up of charge occurs in an explosive environment. Additives are available to minimise the build-up of static; they work by attracting a minute film of water to the surface of the polymer which provides a conductive path for static electricity.

Lightning strikes are an important consideration in aerospace applications.[30] Protection is needed in the form of either a surface film of good electrical and thermal conductivity and a high dielectric strength underlayer, or a resistive surface with a high dielectric strength underlayer with attached metal members.

The insulating properties of plastics, besides causing static problems, give rise to their widespread use in electrical circuitry, such as transformer coil formers and terminal blocks. Any additive in the form of a fibre could be used to modify the conductivity in a desired way as well as being used to improve mechanical properties.

The conductivity of a composite can be increased by the incorporation of a conducting fibre. Figure 1.6 shows the effect on the resistivity of a polymer matrix of the addition of carbon particles.[3] In this Figure, there is a rapid change in resistivity at a carbon concentration of just below 30 phr. Heating of the matrix causes expansion which will increase the distances between the carbon particles, and hence increase the resistivity. Such a system is used as a thermostat: when the desired temperature is reached the resistivity increases, causing a decrease in current such that the heating effect is reduced.

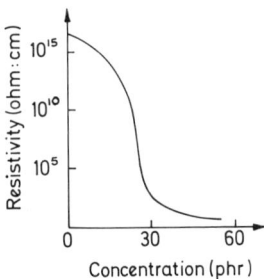

Fig. 1.6. Influence of filler on polymer resistivity in a carbon black/natural rubber system (after ref. 2).

Another area in which thermoplastics containing conducting fibres are useful is the screening of electrical components from electromagnetic interference. One such fibre is made from aluminium coated glass, either in chopped form or in strands of 25 $\mu$m diameter. This is a far cheaper method of obtaining electrical conductivity than using aluminium fibres in solid form.

### 1.4.8. Cost

One unexpected advantage is that in general glass fibre reinforced thermoplastics are cheaper than the matrix material alone. Moreover, because the reinforced material is stiffer, thinner sections will be sufficient and an additional saving in cost may be made.

### 1.4.9. Metal Inserts

Some thermoplastics, polycarbonate being one, do not respond well to having metal inserts moulded into them.[31] After a while cracks propagate around the insert and the moulding loses its strength. Filled grades, both particulate and fibre filled, do not show this undesirable property and can have metal inserts moulded into them.

### 1.4.10. Other Advantages

Glass fibres improve fatigue strength, environmental stress cracking and chemical resistance and make glass reinforced grades attractive materials for under-the-bonnet applications in the motor industry, domestic appliances and other applications where good chemical resistance is required, or where there are harsh environments.

Some of the improvements due to fibre reinforcement have been represented above. Below are some of the disadvantages.

### 1.4.11. Impact Strength

Fibre reinforcement often reduces the impact strength of the composite, particularly when brittle fibres are used in brittle matrices. As would be expected, the impact strength will be very sensitive to the choice of fibre and matrix.

### 1.4.12. Surface Finish and Transmittance

The surfaces of fibre-filled grades are less glossy than their unfilled counterparts. This is particularly apparent when large expanses of dark colour are involved, when some 'flecking' of the fibres beneath the surface may be noticeable. The effect can be reduced by using lighter colours or by the appropriate choice of moulding conditions. A high injection rate will minimise the surface imperfections.[28] Filled grades are invariably opaque or translucent. Transparency is only possible when either the fibres are of a length less than the wavelength of light, in which case their strengthening effect would be small, or the material of the fibre has the same refractive index as the coating and the matrix. This condition would be limited to a small temperature range, unless the temperature coefficient of refractive index of all three materials were the same.[3]

### 1.4.13. Rheological Properties
The viscosity of fibre reinforced grades is higher than that of the matrix material. This effect is greatest at low shear rates and is not particularly important at the shear rates encountered in extrusion and injection moulding, although higher barrel pressures will be needed in these processes. The fibres will also reduce the melt elasticity, which will reduce die swell and may reduce extrudate distortions. The rheology is discussed fully in Chapter 5.

### 1.4.14. Processing
The main problems in processing reinforced grades lie in the increased wear attributable to the presence of the fibres, quartz filled materials being particularly severe on the surfaces of injection moulders and extruders. For this reason, hardened screws and liners, for example ion-nitrided screws and bimetallic liners, are used. Processing and the effects of processing on mechanical properties are discussed in Chapters 5 and 6 and in ref. 3.

## 1.5. VERSATILITY OF FIBRE REINFORCED THERMOPLASTICS

When designing with fibre reinforced thermoplastics, the selection must be made from a great number of different grades because of the multiplicity of possible matrix/fibre combinations. This vast selection gives a designer an opportunity to tailor the material more closely to the application requirements, and often to make products not previously possible for traditional materials. Until design engineers become fully cognisant of the variety and capabilities of these reinforced materials, many of their inherent properties will not be fully utilised.

The microstructure of reinforced thermoplastics is far more complex than that of the matrix material alone, and it is this complexity that renders the variety present. This must be so, because it is the micro-structure that determines the rheological and mechanical properties. In turn, the microstructure can be manipulated during the formulation of the material and during the processing of the product.

In the formation of a fibre reinforced thermoplastic material for a given application, consideration must be given to the choice of matrix material.

The fibre used in a given matrix will have a profound effect on the mechanical properties, and the way in which the fibre interacts with the matrix defines another microstructural variable that will determine the rheological as well as the mechanical behaviour. Fibres used in reinforcement are discussed in Chapter 2 and their effect on the matrix material in Chapter 9. In Chapter 3 the development of natural fibres is discussed. This is an exciting new venture.

The interaction between a fibre and the matrix is usually modified by a coupling agent. Different coupling agents will react in various ways with a given fibre/matrix combination. Chapter 8 deals with coupling agents.

The effect of the processing variables on a matrix material can be profound, giving a variable orientation, and in partly crystalline materials a varying degree of crystallinity. Surface gloss, orientation and built-in strain are all functions of the process variables. The situation with fibre reinforced thermoplastics involving matrix, fibre and coupling agent is, therefore, even more complex. Orientation of the fibres, even more than that of the polymer molecules in the matrix, is of utmost importance, giving mechanical properties that differ along the fibres from those perpendicular to the fibres. For this reason, two sets of data are given, one for the longitudinal and one for the transverse direction. By modifying the process conditions, the degree of orientation can be altered, which may have a profound effect on the mechanical properties of the finished part.

The last variable that is sometimes relevant involves the choice of foaming agent in reinforced microcellular foams. The choice and amount of agent used will partly control the size of the pores, and the process variables will control the thickness of the skin, which has no microcellular structure. The whole subject of structural foams is described by Methven and Dawson in ref. 22.

Table 1.7 shows the effect of short fibres on the mechanical properties of PEEK. The general trend is one of improvement. Only the impact strength as measured by ASTM D256 is reduced considerably, carbon fibre/PEEK being the worst in this respect. The effect of long carbon fibres in PEEK is discussed in Chapter 4.

The prospects for composite materials are improving rapidly and in the high performance sector the growth rate is estimated to be 16% per year with aerospace applications growing at the rate of 22% per year. Traditionally these markets have been dominated by epoxies, which accounted for 80% of resins used in this field in 1984. However, PEEK and polyphenylene sulphide, in particular, are beginning to challenge this dominance in the high performance field, aided by the possibilities of easier and cheaper processing and recyclability.

The majority of reinforced thermoplastics are used in injection mouldings, with glass reinforced polyamides dominating this sector. However, it is possible to reinforce any thermoplastic and important progress has been made recently with the introduction of glass reinforced PET and the development of special coupling agents in the case of glass reinforced polypropylene.

The most recent developments are of great interest and aimed particularly at the automotive industry. In this connection ICI have developed the 'sandwich' moulding process so that it is now possible to injection mould large parts with, for instance, a solid polypropylene high-gloss skin containing a rigid glass reinforced polypropylene core. This approach helps to satisfy many of the conflicting properties required of materials. Another significant development has been the introduction of long-fibre reinforced thermoplastic formulations. These contain glass fibres approximately 10 mm long, compared with less than 0·4 mm found in standard compounds. This is achieved by using the pultrusion process during compounding, giving high impregnation levels and marked improvements in mechanical properties. These materials have been developed by ICI and are currently available under the 'Verton' name.

Ironically, reinforced thermoplastics are, in many cases, not suitable for injection moulding and processing technology developed for thermosets is being increasingly applied to reinforced thermoplastics. For example, filament winding, in modified form, is being used and glass mat thermoplastics (GMTs), based on polypropylene or polyamide and possibly polybutylene terephthalate and polyurethanes, are being developed. These GMTs can be rapidly pressed to shape in modified sheet metal presses. Current applications are in the automotive industry where the main objective is the replacement of metal parts.

It is hoped that the present work will provide a firm basis for study of this area, and that other references (3, 13, 16, 17, 18, 19, 27, 28, 32, 33) will fill in details and aspects not covered in this book.

# REFERENCES

1. Tennant, R. M., *Science Data Book*, Oliver and Boyd, Edinburgh, 1971
2. Brydson, J. A., *Plastics Materials*, Newnes Butterworth, London, 1975.
3. Sheldon, R. P., *Composite Polymeric Materials*, Applied Science Publishers, London and New York, 1982, Ch. 1.
4. *Udel Polysulphone*, Union Carbide (UK) Ltd, Rickmansworth, UK, 1983.
5. *Victrex Polyethersulphone*, ICI PLC, Welwyn Garden City, UK, 1982
6. *Victrex Poly-ether-ether Ketone*, ICI PLC, Welwyn Garden City, UK, 1980.
7. *Deroton Polyester Moulding Compounds*, ICI PLC, Welwyn Garden City, UK, 1983.
8. *Vespel Polyimide*, DuPont (UK) Ltd, Hemel Hempstead, UK, 1984.
9. Pye, A. M., *Materials in Engng*, April 1982, **3**, 407–9.
10. Smith, C. P., ICI PLC, Wilton, private communication.
11. *Noryl Thermoplastics Resins*, General Electric Plastics BV, Bergen Op Zoom, The Netherlands, 1981.
12. Hearle, J. W. S., *Polymers and Their Properties*, Vol. 1, Ellis Horwood, Chichester, UK, 1982.
13. Mascia, L., *Thermoplastics: Engineering Materials*, Applied Science Publishers, London and New York, 1982.
14. Robeson, L. M., Noshay, A., Matzner, M. and Merriam, C. N., *Die Angew Mak. Chem.*, 1973, **29/30**, 47–62.
15. Noshay, A., Matzner, M., Barth, B. P. and Walton, R. K., *Amer. Chem. Soc. Div. Org. Coatings, Plast. Chem. Preprint*, Sept. 1974, 217.
16. Agarwal, B. D. and Broutman, L. J., *Analysis and Performance of Fibre Composites*, John Wiley and Sons, New York, 1980.
17. Piggott, M. R., *Load-bearing Fibre Composites*, Pergamon Press, Oxford, 1980.
18. Pritchard, G. (Ed.), *Developments in Reinforced Plastics — 1*, Applied Science Publishers, London and New York, 1980.
19. Hancox, N. L. (Ed.), *Fibre Composite Hybrid Materials*, Applied Science Publishers, London and New York, 1981.
20. Crawford, R. J., *Plastics Engineering*, Pergamon Press, Oxford, 1981, Ch. 2.
21. Maxwell, J., *J. Plast. Rubber Int.*, 1983, **8**(2), 45–49.
22. Hilyard, N. C. (Ed.), *Mechanics of Cellular Plastics*, Applied Science Publishers, London and New York, 1981.
23. Monte, S. J. and Sugarman, G., in *Additives for Plastics*, Vol. 1, R. B. Seymour (Ed.), Academic Press, New York, 1978.
24. Plueddemann, E. P., *Silane Coupling Agents*, Plenum Press, New York, 1982.
25. Plueddemann, E. P. (Ed.), *Interfaces in Polymer Matrix Composites*, Academic Press, New York, 1974.
26. Schmitt, G. F., Jr, *SAMPE J.*, July/Aug. 1977, 16–22.
27. Pritchard, G. in *Fibre Composite Hybrid Materials*, N. L. Hancox (Ed.), Applied Science Publishers, London and New York, 1981.
28. McGregory, R. C. in *Developments in Injection Moulding — 1*, A. Whelan and J. L. Croft (Eds), Applied Science Publishers, London and New York, 1978.
29. Wilson, E. L. in *Flame Retardancy of Polymer Materials*, Vol. 3, W. C. Kuryla *et al.* (Eds), Marcel Dekker, New York, 1975, pp. 254–334.

30. Clark, H. T., ASTM STP 546, 1974, 324–42.
31. *Lexan Polycarbonate Resins,* Engineering Polymers Ltd, Wilmslow, UK, private communication, 1984.
32. Folkes, M. J., *Short-fibre Reinforced Thermoplastics,* John Wiley and Sons, New York, 1982.
33. Titow, W. V. and Lanham, B. J., *Reinforced Thermoplastics,* Applied Science Publishers, London and New York, 1975.

*Chapter 2*

# Fibres, Whiskers and Flakes for Composite Applications

H. W. Rayson, G. C. McGrath

*Department of Metals and Materials Engineering*

and

A. A. Collyer

*Department of Applied Physics, Sheffield City Polytechnic, Sheffield, UK*

## 2.1. INTRODUCTION

In this chapter it is intended to examine the mechanical properties of some of the more widely used fibres in short fibre reinforced thermoplastics and to present in brief form the well-known theories associated with both short fibre and flake/ribbon reinforcements. It is hoped that this chapter together with the comprehensive list of references will give a thorough introduction to the subject of short fibre reinforced thermoplastics.

## 2.2. FIBRES AND WHISKERS FOR COMPOSITE APPLICATIONS

### 2.2.1. Manufacture of Fibres and Whiskers

Fibre technology covers a broad spectrum of science and engineering, and includes all aspects from fibre development to composite design, fabrication and utilization. A schematic diagram (Fig. 2.1) illustrates the scope of this technology and indicates that the study of composite materials is truly an interdisciplinary one.

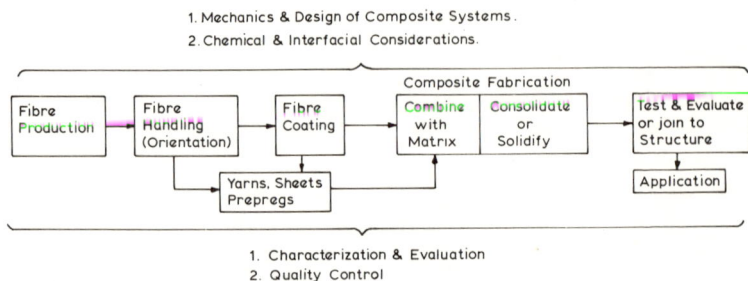

**Fig. 2.1.** Aspects of fibre technology.

The manufacturing techniques employed for obtaining the reinforcements are many and varied[1,2] and even as early as 1924 a method of drawing fine metallic filaments had been developed.[3] It is beyond the scope of this survey to describe in detail the many processes used, and only an indication of the methods available for fibre and whisker manufacture will be given. Some of the methods used are as follows.

(i) *Whisker growth*: used for the production of, for example, alumina, silicon nitride and graphite. Furthermore, (cubic) silicon carbide whiskers are being developed in bulk quantities.[4]

(ii) *Glass drawing*: glass and also polycrystalline fibres such as alumina are being produced by this method.

(iii) *Wire drawing*: used for metal wires, e.g. tungsten and molybdenum.

(iv) *Vapour deposition*: used for multiphase fibres such as boron in tungsten.

(v) *Precursor method*: carbon fibres are produced by this method, which is discussed in Section 2.3.4.

### 2.2.2. Mechanical Properties of Fibres and Whiskers

Table 2.1 lists the general types of reinforcement that are available. It is important to characterize adequately the mechanical properties of the fibres for several reasons.

(1) To provide a basis for assessing their potential reinforcing strength.

(2) To serve as a quality control check on the forming process.

(3) To provide a measure of the scatter in strength.

(4) To use the data as a tool for evaluating the damaging effects of processing environment and composite fabrication on the fibres.

The test methods (e.g. tensile tests, bend tests, whisker testing, etc.) used for fibres and whiskers and the inherent drawbacks of these methods have been comprehensively reviewed.[5]

Most high-strength fibres are brittle, and therefore they are sensitive to surface flaws and other defects. Zweben[6] has considered the importance of the statistical scatter in fibre strength when the fibres are tested at constant gauge length, and has also illustrated the sensitivity of fibres to flaws by the observed decrease in mean strength as the gauge length is increased.

Coleman[7] has emphasized that the ultimate tensile strength of a bundle of fibres is usually less than the mean strength of the fibres tested individually. He has plotted (Fig. 2.2) the ratio of bundle strength $\sigma_B$ to mean fibre strength $\sigma_f$ against the coefficient of variation $\bar{c}$, which is defined as the ratio of the standard deviation to the mean fibre strength. Typically $\bar{c}$ can be expected to be between 0·08 to 0·25.

A theory has been derived for predicting the strengths of

TABLE 2.1
Properties of Fibres

| Fibre | Melting or softening point (°C) | Density, $\sigma$ (kg m$^{-3}$) | Tensile strength, $\rho$ (MN m$^{-2}$) | Specific strength, $\rho/\sigma$ (MN m kg$^{-1}$) | Young's modulus, E (GN m$^{-2}$) | Specific modulus, E/$\rho$ (m Nm kg$^{-1}$) |
|---|---|---|---|---|---|---|
| *Whiskers* | | | | | | |
| Ceramic | | | | | | |
| Al$_2$O$_3$ | 2 038 | 3 961 | 4 136–24 132 | 1·044–6·092 | 413·7–1 034 | 104·4–261·0 |
| BeO | 2 571 | 2 853 | 13 100 | 4·592 | 344·7 | 120·8 |
| B$_4$C | 2 449 | 2 521 | 13 789 | 5·470 | 482·6 | 191·4 |
| SiC | 2 688 | 3 213 | 13 789–41 369 | 4·292–12·87 | 482·6–1 034 | 150·2–321·8 |
| Si$_3$N$_4$ | 1 899 | 3 186 | 4 826–13 789 | 1·514–4·328 | 275·8–379·2 | 86·57–119·0 |
| Graphite | 3 649 | 1 662 | 19 616 | 11·80 | 703·3 | 423·2 |
| Metal | | | | | | |
| Cr | 1 888 | 7 202 | 8 894 | 1·235 | 241·3 | 33·50 |
| Cu | 1 082 | 8 919 | 2 944 | 0·330 1 | 124·1 | 13·91 |
| Fe | 1 534 | 7 839 | 13 100 | 1·671 | 199·9 | 25·50 |
| Ni | 1 454 | 8 975 | 3 861 | 0·430 2 | 213·7 | 23·81 |

*Filaments*

| | | | | | |
|---|---|---|---|---|---|
| Glass | | | | | |
| Type-E | 699 | 2 548 | 1·353 | 72·39 | 28·41 |
| Type-s | 838 | 2 493 | 1·797 | 86·87 | 34·84 |
| Type 4H-1 | 899 | 2 659 | 1·893 | 99·97 | 37·60 |
| SiO$_2$ | 1 696 | 2 188 | 2·678 | 72·39 | 33·08 |
| Polycrystalline | | | | | |
| Al$_2$O$_3$ | 2 038 | 3 158 | 0·654 8 | 172·4 | 54·59 |
| ZrO$_2$ | 2 649 | 4 847 | 0·426 7 | 344·7 | 71·12 |
| BN | 2 982 | 1 911 | 0·721 6 | 89·63 | 46·90 |
| Graphite | 3 649 | 1 413–1 994 | 0·878 2–1·296 | 172·4–413·7 | 122·0–207·5 |
| Multiphase | | | | | |
| B | 2 299 | 2 631 | 1·048 | 379·2 | 144·1 |
| B$_3$C | 2 449 | 2 354 | 0·966 4 | 482·6 | 205·0 |
| SiC | 2 688 | 3 463–3 518 | 0·696 8 | 413·7–461·9 | 119·5–131·5 |
| SiC on B | 2 299 | ~2 770 | ~0·995 7 | 379·2 | ~136·9 |
| TiB$_2$ | 2 982 | 4 487 | 0·023 0 | 510·2 | 113·7 |
| Metal | | | | | |
| W | 3 390 | 19 310 | 0·207 1 | 406·8 | 21·07 |
| Mo | 2 621 | 10 221 | 0·215 8 | 358·5 | 35·07 |
| Steel | 1 399 | 7 756 | 0·533 3 | 199·9 | 25·77 |
| Be | 1 282 | 1 828 | 0·697 5 | 241·3 | 132·0 |

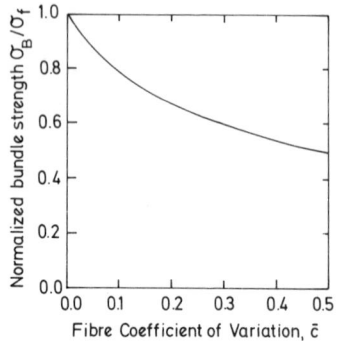

**Fig. 2.2.** Relationship between the fibre coefficients of strength variation and the normalized strengths of fibre bundles.

discontinuous-fibre composites when the fibres are flawed to varying degrees.[8] An important conclusion is that if a composite contains badly flawed fibres, then these fibres can never contribute more than 50% of their maximum inherent strength to the strength of the composite. This is in contrast to the prediction that flawless discontinuous fibres can contribute a maximum of $\frac{6}{7}$ of their strength to the composite strength.[9]

## 2.3. TYPES OF FIBRES WIDELY USED FOR REINFORCEMENT OF POLYMERS

### 2.3.1. Asbestos

Asbestos is a strong, crystalline, inorganic, naturally occurring fibrous mineral, with a unique combination of high strength, high elastic modulus, thermal stability, low thermal conductivity, good wear resistance, and low density, at a moderate price. The properties are summarized in Table 2.2. Chemically, asbestos is a group of hydrated metallic silicates, divided into two classes; the first consists of chrysotile, hydrated magnesium silicate, $Mg_6[(OH)_4Si_2O_5]_2$, which makes up the vast majority of asbestos production (i.e. over 95%). There are five types of asbestos in the second class, the most important member being crocidolite, or blue asbestos, $Na_2MgFe_5[(OH)Si_4O_{11}]_2$, noted for its acid resisting properties, although mainly used in asbestos-cement products. It represents a health hazard, like other grades, unless treated very carefully.

Asbestos is graded according to the mean fibre length, from grade 2

TABLE 2.2
Physical and Mechanical Properties of Asbestos

| Property | Chrysotile | Crocidolite | Amosite | Anthophyllite |
|---|---|---|---|---|
| Colour | White to grey | Blue | Brown | Brown to grey |
| Tensile strength, MN m$^{-2}$ | 2 068 | 3 447 | 1 103 | 2 413 |
| Modulus of elasticity, GNm$^{-2}$ | 159·9 | 186·8 | 162·7 | 155·1 |
| Hardness, Mohs | 2·5–4·0 | 4·0 | 5·5–6·0 | 5·5–6·0 |
| Flexibility | Good | Fair | Poor | Poor |
| Specific gravity | 2·4–2·6 | 3·2–3·3 | 3·1–3·2 | 2·9–3·2 |
| Specific heat J kg$^{-1}$ °C$^{-1}$ | 1 114 | 879·2 | 808·1 | 879·2 |
| pH | 10·3 | 9·1 | 9·1 | 9·4 |
| Refractive index | 1·50–1·55 | 1·70 | 1·64 | 1·61 |
| Fibril diameter, nm | 160–300 | 600–900 | 600–900 | 600–900 |
| Coefficient of cubical expansion, °C$^{-1}$ | 9 × 10$^{-5}$ | — | — | — |

(mean length 1·59 cm) to grade 7 (mean length 0·079 cm). Separation into grades is facilitated at the mine since the large fibres, from the wider veins, separate from the rock more readily. Some blending is also carried out in mixers, to give a required fibre mean size and size range.

*Chemical Properties of Chrysotile*
This predominant grade of asbestos has a layered structure of silica tetrahedra and magnesium hydroxide, the latter being readily attacked by concentrated acids (blue asbestos is more acid resistant). Chrysotile is more resistant to caustic alkalis than other grades. Its general resistance to other chemicals is demonstrated by its lack of reactivity in a wide range of hostile environments. It is particularly inert in standard atmospheric conditions. Although, like many polymers, it reversibly absorbs water (up to 3% in high humidity situations), this has little effect on its composite-forming behaviour.

*Physical and Mechanical Properties*
The important mechanical and physical properties of chrysotile are given[10] in Table 2.2. Up to 60% or more of asbestos is frequently incorporated into short fibre composites, so many of these properties are mirrored in the overall properties of the composites. Mechanical

properties such as tensile strength and Young'smodulus are difficult to measure on short, small-diameter fibres. The small fibre diameters (200 nm) ensure high aspect ratios, even in the shorter lengths of fibres. The strengths and moduli are very high, exceeding those of high-strength steels, for example, by about 50%. Hardness is not excessive ($\sim 4.0$ Mohs), which partly explains the usage of asbestos in brake pads and linings. Flexibility is particularly high for chrysotile, and although excessive processing may appear to disrupt the fibres, there is little loss of length, the main effect of bending being separation of fibrils from fibre bundles, with an apparent decrease in fibre diameter. This flexibility in bending would also be expected to be exhibited when the fibres are incorporated in a relatively brittle matrix type of composite. Unfortunately the impact strength in an asbestos composite is frequently disappointing, often being less than that of the pure polymer matrix, as shown in ABS, polyethylene and other matrices,[11] where it is clear that the fibre–matrix interface must be too weak to give much energy absorption on fracture, without careful production in the presence of coupling agents. An alternative explanation of low impact strength could be easy crack propagation between individual members of asbestos fibre bundles, which would probably still occur even if effective coupling agents were developed.

*Enhancement of Properties in Polymer Matrices*
Asbestos, although denser than polymers, maintains the low density advantage of polymers reasonably well compared with metals, and may reduce costs because of section reductions permitted by the increased flexural strength and other enhanced mechanical properties. The chemical stability of the polymers will also be maintained, in general, in the composites. There are a number of improvements in resistance to temperature effects, i.e. lower thermal expansion (increased dimensional stability), better heat deflection, improved high-temperature strength, and higher creep strength. Improvements can also be expected in most of the mechanical properties in Table 2.2. Flow control is improved in moulding operations (due to an effective viscosity increase), there is negligible fibre degeneration in forming operations, and a good surface finish is generally obtained.

Disadvantages include disappointing impact strength (frequently a main advantage of fibre reinforcement), loss of transparency, some increase in specific gravity, and a lack of thermal stability which is ascribed to the high surface area of the fibres. Thermal stability is particularly bad for chrysotile in polypropylene, partly due to oxidation. Inhibitors such as phenol derivatives have a beneficial effect on stability in epoxy resins.

A wide range of polymers is reinforced with asbestos in injection moulded thermoplastic products, giving increased strength and stiffness, with reduction in heat distortion and in shrinkage. Although asbestos is frequently incorporated in polypropylene, there are thermal degradation problems, as previously mentioned (PVC is similarly affected). A number of car components (e.g. Rover glove boxes) are made from asbestos reinforced polypropylene, because of its flame retardant properties. Owing to increased stiffness at elevated temperatures such products have the economic advantage of reduced cycle times during moulding, since they can be withdrawn from the mould at higher temperatures. Other applications of this material include heater housings, radiator fan shrouds, automotive lamp housings and central heating components.[12] It is interesting to note that talc filled polypropylene can give similar flame retardant advantages.

*Applications*
Calendered sheet based on PVC is widely used for flooring. The incorporated fibres may merely be used as a filler, or give reinforcement because of preferred orientation in certain directions in the plane of the sheet.

*Health Hazards*
After long exposure to dusty atmospheres containing asbestos (particularly crocidolite, or 'blue asbestos'), asbestosis, bronchial cancer and mesothelioma may develop. It has been stated in official reports that occupational exposures much higher than those met by the general public do not give an increased risk of lung cancer. Nevertheless, permitted occupational exposures have been lowered in recent years, the threshold in the USA being reduced in 1976 to two fibres per millilitre. Control limits recommended in the UK by the Advisory Committee on Asbestos, with effect from January 1983, were 0·2 fibre/ml, when measured as a time-weighted average over any four-hour period, for dust consisting of or containing any crocidolite. Presumably such levels can be reached by careful control of the atmosphere in the asbestos industry.

*Alternatives to Asbestos*

Much effort is being put into the replacement of asbestos in brake linings, in which there is the possibility of generation of dust during application. This is an area of commercial confidentiality, in which existing and potential material compositions are not readily revealed. There is also extensive patents literature on this subject. Substitution for asbestos is not simple, due to its unique combination of properties. A low-cost, inert filler could apparently replace asbestos, but higher-cost reinforcing fibres would be needed to compensate for some of the advantageous properties of asbestos that would otherwise be lost. Possible substitutions include glass, steel and Kevlar (aramid) fibre. All three have high strength and elastic modulus. Kevlar and steel have the required thermal stability. Good quality asbestos linings may contain 50–60% asbestos, 15% $BaSO_4$, 25% organics and 5% of other materials. The excellent friction, wear and fade performance of this material can apparently be matched by incorporating 5% of 12 $\mu$m-diameter Kevlar fibre, with 50% dolomite, 15% $BaSO_4$, 15% cashew friction particles and 15% NC 126 Resin (a product of the 3M Company).[13]

### 2.3.2. Kevlar

*Introduction*

Kevlar is a DuPont trademark for poly($p$-phenylenediamine terephthalamide). It is an aromatic polyamide (aramid) fibre which has the highest specific strength of any commercially available fibre. (Nylon fibres, which also have excellent properties, are analogous, being aliphatic polyamides.) Other properties which make Kevlar such an outstanding material include its low density, ability to retain its major properties up to 180°C, lack of embrittlement at low temperatures (to −196°C), resistance to chemicals and ultraviolet radiation (except at the surface), and stretch resistance (4% compared with 19% for nylon).

Kevlar is considered the most important man-made fibre to be discovered since nylon. First discovered by Stephanie Kwolek in 1965, Kevlar was then improved over an eight-year period; the main application for which it was aimed was radial car tyres. Now two main grades are available, Kevlar 29 (for ropes, cables, friction materials and bullet-proof personnel protection), and the higher modulus Kevlar 49, used in reinforced composite boat hulls, aircraft, missile structures and racing-car bodies. A summary of the more important properties of

Kevlar in comparison with nylon, E-glass and steel is given in Table 2.3.

TABLE 2.3
Properties of Kevlar and Competitive Materials

|  | Nylon | Kevlar 29 | E-Glass | Kevlar 49 | Steel |
|---|---|---|---|---|---|
| Specific gravity | 1·14 | 1·44 | 2·55 | 1·45 | 7·86 |
| Tensile strength, kN m$^{-2}$ | 999 | 2 758 | 1 517 | 2 758 | 1 965 |
| Tensile modulus, MN m$^{-2}$ | 5·52 | 82·74 | 68·95 | 131·0 | 200 |
| Elongation, % | 18 | 5 | 3 | 2·4 | 2 |

## Recent Investigations of the Properties and Behaviour of Kevlar

*Fibre morphology.* The morphology of Kevlar fibres is very relevant to their mechanical behaviour. Techniques such as electron microscopy, electron diffraction, ultramicrotomy and ion thinning[14] have revealed the longitudinal and transverse microstructures. An outer 'skin' is seen, enclosing the 'core' of the fibre, a layered structure of rod-like crystallites with their major axes parallel to the fibre axis. The layers of crystals are also parallel to the fibre. Hydrogen bonds in the structure are radially oriented. Raman spectroscopy[15] reveals that, under stress, crystal alignment is further improved, accompanied by an opening of the angles in the amide linkage of the polymer chain. Further studies[16] reveal that the crystal layers parallel to the fibre axis are regularly pleated, and thus arranged radially.

*Behaviour in compression.* Buckling and other forms of instability under compressive stresses parallel to the fibre axis are a feature of all objects with a large aspect ratio (height/diameter). In fibre bundles, or fibrillar structures, this effect may be intensified by weak fibre–fibre or fibre–matrix interfaces within a composite material. The occurrence of internal instability at low stresses observed in an epoxy matrix has been explained in terms of the high mechanical anisotropy of the fibres. Failure regions show a high degree of interfacial debonding, accompanied by fibrillation, i.e. formation of bundles of very small diameter fibrils,[17] both features accounting for disappointing compressive strengths. Plastic deformation is observed in the fibres at the low compressive stresses which cause composite fracture before there is any plastic deformation of the matrix.[18]

*Environmental behaviour.* Effects of many chemicals on the strength of Kevlar 29 and Kevlar 49 have been studied. Under some circumstances industrial solvents, such as acetone and ethyl alcohol, have no effect. The effects of lubricants, water and fuels are frequently insignificant (although water is absorbed reversibly). Very strong acids and bases do, however, give appreciable strength losses, although Kevlar 49 is more resistant than glass to HF and NaOH.

The effects of ultraviolet light are minimized in ropes, compared with single yarns, because the self-shielding effect limits the degradation to the surface layers of the material. Typically, without protection, strength losses of 30% are observed.

*Temperature effects.* Kevlar has outstanding properties at elevated temperatures, compared with other organic fibres, showing no melting point but decomposing at about 400°C. Deterioration is mainly due to oxidation. Long times of exposure cause degradation at temperatures well below the decomposition temperature so that 180°C is generally reckoned to be the maximum effective working temperature.

*Fatigue properties.* Fatigue properties of Kevlar 29 have been shown to be superior to those of glass and carbon fibres in high-strain flexing and buckling tests,[19] because the Kevlar fibres could survive by absorbing the stresses by plastic yielding in compression. Fracture studies confirmed that Kevlar fibres have low strengths perpendicular to the fibre axes, since the fibres split axially under compressive loads.[19, 20] As is commonly observed in other materials, fatigue life depends on stress amplitude as well as on the maximum applied load.[21]

In composites, Kevlar 49 reinforcement, compared with glass fibre reinforcement, is claimed to confer superior fatigue resistance, with less propagation of damage and higher vibrational damping.[22]

*Impact strength.* Impact properties of Kevlar are favourable and lead to its use in protective clothing, helmets, etc. Tests on hybrid laminates containing a combination of carbon and Kevlar fibres show that properties after impact damage are better than those obtained with one type of fibre only.

*Creep.* Creep properties of Kevlar 49 have been measured and compared with properties of Kevlar/epoxy composites[23] (recent work on composites has very much favoured the choice of an epoxy matrix). In

1000 h room temperature tests, only transient creep (i.e. diminishing rate vs time) was observed in the composites. The matrix is more creep resistant than the fibres at low stresses, the situation reversing at higher stresses. Twisted Kevlar 49 yarn shows very low steady-state creep rates when loaded to 50% of its ultimate tensile stress at room temperature. Stress rupture behaviour at comparable loadings in terms of percentage of tensile strength show that Kevlar 49 is highly superior to nylon 66 at room temperature.

*Kevlar–matrix interface effects.* In attempts to improve composite properties, studies have been made of Kevlar–matrix interfaces, in common with those of other composites. For example, functional groups have been provided on Kevlar fibre surfaces[24] (using the metallation reaction in dimethyl sulphoxide) incorporated in a polyethylene matrix, and fracture surfaces then studied by scanning electron microscopy (SEM).

Enhancement of adhesion to polystyrene has been obtained by *in situ* bulk polymerization of styrene, using divinylbenzene as crosslinking agent. The need for interface strengthening is illustrated by experiments to measure bond strength at the interface in the popular epoxy/Kevlar combination,[25] when wide variability was obtained in the observed bond load of single filaments during pull-out tests. After debonding a subsequent lower frictional load was observed.

*Short Kevlar fibres.* Like glass and carbon fibres, Kevlar is normally obtained in continuous form, and there are, unfortunately, some problems in obtaining shorter, damage-free 'chopped' fibres suitable for use in plastics moulding, from the longer lengths. These problems are partly due to the other beneficial properties of the fibres. Chopper guns used for glass fibre cutting are generally ineffective with Kevlar. Tungsten carbide gritted tools have been recommended, as have water jet cutters.[26] The problem of hole drilling in Kevlar composites has also received attention.[27]

*Applications of Kevlar.* Widespread usage, with an annual growth rate of 20%, applies both in fibre and composite form, the latter mainly with epoxy resin matrices. The main areas of development are in tyres, aircraft, marine and automobile engineering (low density is of particular value in transportation where it increases the effective payload), ropes for many industries, parachutes, pressure vessels (often rocket bodies), sports goods (competing with, or enhancing, the effect of carbon fibres in shafts, etc.), brake linings (as replacement for asbestos in combination with weaker, inert material) and protective clothing. Avtek Corporation have developed a high-performance business jet aircraft made almost entirely of aramid composites,[28] giving significant weight reduction compared with more traditional materials. Specific applications in the aircraft industry include airframe components, undercarriage fairings, moulded airship shell (the Skyship 500),[29] replacement of aluminium in leading and trailing edges, tailfin and rudder assemblies. A similar complexity of applications is developing in the marine and automotive industries, with emphasis on mouldable, lightweight rigid hulls, body panels, etc.

### 2.3.3 Glass

As a continuous or short fibre reinforcement medium, glass fibre has many advantages, including availability, well-established technology, low cost, reliability, and resistance to chemical and environmental degradation. One limitation of glass reinforced plastics (GRP) in general, and the sheet moulding compounds (SMC) more recently developed, is the relative lack of stiffness. A recent investigation of mass fraction of fibre up to about 65% in a thermosetting polyester resin demonstrates that the stiffening effect of glass fibre is exceeded by that of Kevlar, low-modulus carbon and high-modulus carbon, in order of increasing effectiveness.[30] This is the outcome of the low modulus of glass, compared with the other fibres. The density of glass is also higher than those of carbon and Kevlar, composites of which therefore have a density advantage over GRP.

Glass is normally an amorphous material, based on silica. The simplest form chemically, with the most regular crystal structure, is fused silica, which has not been widely used in fibre reinforced thermoplastics. The high melting point and excellent high-temperature properties of silica are useful in materials for high-temperature applications. Silica precoated with aluminium has been studied in detail as a model reinforced material. Although stiffening at room temperature is negligible, due to similarity in the elastic moduli of $SiO_2$ and Al, there is an increase in high-temperature rigidity due to the lower temperature coefficient of the elastic (Young's) modulus of $SiO_2$ compared with that of Al.

Types of glass widely used in the plastics industry generally contain between 50 and 65% $SiO_2$, the other oxides present being $Al_2O_3$, CaO, MgO, $Na_2O$, $B_2O_3$ and BaO.

Since density is important in many applications of fibre reinforced plastics, the selection of a low density grade such as D-glass might be considered (S.G. of 2·16 compared with 2·54 for E-glass). However, in terms of strength or modulus per unit mass, D-glass shows no advantage, but it has more favourable dielectric properties. The compositions of D- and E-glass are given in Table 2.4.

TABLE 2.4
D- and E-Fibre Glass Compositions

| | *Fibre glass composition* | |
|---|---|---|
| *Property* | *D* | *E* |
| Description | Glass with improved dielectric strength and low density, developed for improved electrical performance | Borosilicate type used for major share of all reinforcement applications. |
| Chemical composition(%) | | |
| $SiO_2$ | 74·5 | 54·0 |
| $Al_2O_3$ | 0·3 | 14·0 |
| $Fe_2O_3$ | — | 0·2 |
| CaO | 0·5 | 17·5 |
| MgO | — | 4·5 |
| $B_2O_3$ | 22·0 | 8·0 |
| $Na_2O$ | 1·0 | 0·6 |
| $K_2O$ | 1·5 | — |
| $ZrO_2$ | $(Li_2O–0·5)$ | — |
| $SO_3$ | — | — |
| $F_2$ | — | 0·1 |

Low thermal expansion coefficients may be an advantage in maintaining thermal dimensional stability. Large differences between expansion rates of fibre and composite may lead to changing residual stress patterns, and possible distortion, in components subjected to temperature changes in service.

Optical properties such as transparency are improved by using resins with similar refractive indices to glass. The optical properties of glass are improved by lowering the iron oxide ($Fe_2O_3$) content.

*Glass Fibre Production Methods*

In the direct melt process, the glass is melted in a furnace and is forced by gravity into a bushing at the base of the furnace forehearth. The bushing contains a perforated baffle plate which improves homogeneity. The glass is then cooled by entering the bushing tips, which are exposed to the surrounding air. The tips are generally composed of multiples of 204 single filaments arising by drawing down from each tip on to a high-speed drum. The filament has a diameter dependent on the velocity of the drawing. The filaments from each bushing are collected and size is applied at a collecting point.

Various finishes may be applied to the fibres during manufacture to improve subsequent bonding with specific matrices, e.g. sizing containing chrome, silane or titanates as described in Chapters 1 and 8.

### 2.3.4. Carbon Fibres

*Manufacture, Mechanical Properties and Microstructure*

The technology of carbon fibre manufacture[31-34] has been continually refined resulting in enhanced mechanical properties,[35-37] and the detailed nature of the fibre structure has been frequently investigated.[37-39]

Many precursors, such as inexpensive pitch,[40] phenolhexamine polymers[41] and a variety of acrylics,[42] have been evaluated for fibre production. At present cellulose (viscose rayon) or acrylic fibres (Courtelle — polyacrylonitrile based) are used commercially as precursors for most American fibres, whilst acrylic fibre is the basis of the British product.

The American process essentially consists of a slow pyrolysing treatment at 1000 °C in an inert atmosphere, followed by hot-stretching at temperatures in excess of 2000 °C. The latter treatment increases both strength and modulus. Briefly, the British process involves low-temperature oxidation at 300 °C whilst the fibres are under tension, carbonization at 1000 °C and finally high-temperature heat treatment in an inert atmosphere. During the initial oxidation treatment carbon atoms become orientated. The final heat treatment temperature determines the type of fibre produced. The strengths of carbon fibres go through a maximum at a final treatment temperature of about 1500 °C whereas their elastic moduli continue to increase. This behaviour gives rise to two types of fibre: Type I, heated to 2600 °C and having the higher modulus (380–448 GPa) and a somewhat lower ultimate tensile strength

(1330–2070 MPa); and Type II, heated to approximately 1500 °C, with the higher strength (2410–3100 MPa) and a somewhat lower modulus (241–310 GPa). The inverse relationship between modulus and strength can be modified by high-temperature stretching as practised in the USA. Cellulose, as used in the USA, gives rise to a non-graphitizing type whilst the British precursor gives a graphitizing type of carbon.

Each filament produced, of a diameter of 7–9 $\mu$m, consists of a number of fibrils each of which in turn is made up of carbon atoms arranged in a hexagonal crystal lattice oriented with the $c$-axes of the crystals almost normal to the major axis of the fibre. These crystal lattices are turbostratically deployed. The structure of the poly-crystalline fibres is based on the graphite lattice in which the atoms in each layer plane are 0·142 nm apart and are covalently bonded whilst the distance between one layer and the next is 0·335 nm. Weak van der Waals type bonding occurs between the layers. The magnitude of the assembly of crystallites is controlled by the degree of graphitization of the fibre, graphite crystallites (50–100 nm) being larger than carbon crystallites (5–10 nm). There is a strong resemblance between the structural layout of graphite and Kevlar fibres.

The structure of high-strength graphite whiskers[32] has been proposed as consisting of one or more concentric tubes, each tube being in the form of a scroll or rolled-up sheet of graphite layers, extending continuously along the length of the whisker, with the $c$-axis perpendicular to the whisker axis.

Well-developed graphitic fibre surfaces give rise to low values in interlaminar shear strengths when embedded in a resin matrix. This drawback can be overcome by whiskerizing[43] the fibre surfaces but, alternatively, surface treated fibres are available for enhancing fibre-matrix bond strengths. The decrease in the shear strength of a plastic matrix composite with increasing fibre modulus has been investigated.[39] Characterization of the fibre surface using polarized light, electron and X-ray diffraction techniques indicated that the surface crystal structure might be different from that of the bulk fibre, and thus the surface might possess different mechanical properties. It was found that surface crystallites were more aligned than core crystallites and that this might cause shear at the fibre surface. Additionally, the fibre surface energy decreases as crystallites become more parallel to the fibre surface. Consequently, it was concluded that as the surface preferred orientation increased (increased modulus), matrix wetting and adhesion propensity decreased together with the shear strength of the fibre surface itself.

Thus the fibre surface and chemical interface both decrease in shear strength as the fibre modulus increases.

The mechanical properties[35, 37, 44-48] of carbon fibres have been widely investigated. The addition of boron to PAN based fibres has increased their stiffness whilst their breaking stress appeared to be unaffected; neutron irradiation increased both properties.[37] It was suggested that hot-stretching could be used to eliminate dislocations piled up at crystallite boundaries that otherwise could have formed undesirable stress concentrations. Other mechanisms, such as solid-solution hardening, were also suggested as a means of improving mechanical properties. In analysing the bending behaviour of graphitized rayon filaments,[44] using loop tests, some evidence of inelastic behaviour was observed. The results were interpreted in terms of rupture of crosslink bonds between adjacent fibrils, using stranded wire for analogy, and it was suggested that it is the fibrillar microstructure which sets a limit to the intrinsic strength of the fibres. Both fibre strength and modulus have been found[45-47] to increase with decreasing fibre diameter whilst these properties do not appear to be dependent on strain rate.[47] A study by Moreton[48] has shown short fibres to be stronger than longer ones. This strength/length effect has been attributed to the frequency of flaws scattered along fibre lengths. The defect type and defect distribution were not influenced by the final heat treatment temperature.

The simultaneous increase in UTS (ultimate tensile strength) and Young's modulus with decreasing diameter has been ascribed to variation in fibre structure across the diameter.[40-47] A 'sheath–core' relationship has been postulated[45] in which crystallites close to the fibre surface tend to be larger and better aligned (i.e. the crystal *a*-axes are almost parallel to the fibre axis) than are crystallites in the fibre core. Thin fibres contain a higher proportion of well aligned crystallites than do thicker ones — hence the increase in fibre modulus with smaller diameter fibres. At the same time, there is a greater proportion of flaws (basal plane cracks) in the fibre core than in the sheath, again giving rise to higher values of UTS for thinner fibres. This sheath–core effect is found to be more pronounced than the strength/length effect when considering equal volumes of fibre. Kawamura and Jenkins[41] have also observed a similar diameter dependence of UTS and modulus for their glassy carbon fibres.

*Electron Microscopy*
Electron microscopy has been extensively used in an effort to determine

accurately the longitudinal[40, 49-52] and cross-sectional[53] structure developed in fibres by various heat treatments Johnson and Tyson,[49] using microtomed slices, have indicated that a longitudinal section of graphitized fibre confirms the impression of a fibrillar structure with crystallite regions separated by sharp-edged elongated voids having their basal planes parallel to the fibre axis. Peripheral regions of cellulose-based fibres have also been examined[49] and again the presence of highly-oriented crystallites, together with micropores, have been detected.

The aforementioned microstructures of fibres were revealed after progressively higher graphitizing temperatures had been employed. Microvoid elongation with increasing graphitizing temperature was attributed to a decrease in the density of branching points along crystallite ribbons.

The formation has been observed[52] of three-dimensional graphite platelets which bear little relation to the original fibre microstructure. An iron–silicon solid solution, formed during heat treatment, was thought to catalyse the growth of these platelets. Lengths of fibre containing such platelets exhibited lower fracture strengths than did uncontaminated fibres. Similar platelets have also been observed in undoped fibres and were attributed to the presence of segregated trace impurities.[51]

Examination of cross-sectioned fibres[53] reveals a twisting complex arrangement of carbon layers going in every direction, in contrast to the long well-ordered planes seen longitudinally. Severe folding and doubling of crystallite bands is observed with no apparent order between them.

## X-ray Studies

The application of X-ray techniques, augmented by electron microscopy, has proved successful in accurately defining the fine structure of fibres. An end-to-end stacked, turbostratic crystallite model, misoriented so that tilt and twist boundaries are formed between the crystallites has been proposed.[49] The individual crystallites are separated by grain boundaries and narrow voids. Such a model has been described elsewhere[37] and other X-ray evidence[31] is consistent with this description.

Perret and Ruland[54] have studied the pore structure in detail by small-angle X-ray scattering. The pore walls were found to consist of turbostratic graphite with layer planes lying nearly parallel to the fibre axis. Evidence for the development of such porosity in graphitized polyacrylonitrile-based fibres, as observed by an increasing surface area, has been found during vacuum heat treatment below 1000°C.

The Debye–Scherrer technique has revealed marked changes in (002) diffraction arcs as a result of impurity concentrations[52] and also the high degree of preferred orientation obtained by 'stretch-graphitizing' pitch precursor fibres.[40]

Graphite fibre surfaces have been characterized by Raman spectroscopy,[55] which examines only thin surface layers of sample due to the high extinction coefficient of the graphite for the laser beam. It was shown that surface crystallites were larger than those in the bulk fibre. Consequently, a weak surface layer was postulated which could adversely affect composite shear strengths.

## 2.4. THEORY OF SHORT FIBRE REINFORCEMENT

The advantages of fibre reinforcement and the types of fibres used in reinforced thermoplastics have been described, and now a summary of the theory is discussed.

The prediction of mechanical properties of aligned, long fibres is difficult to achieve theoretically and further complications exist in dealing with short fibre reinforcement. This is due to the possibility of a spectrum of fibre lengths and orientations caused by processing into the final part. Even if these two possibilities are discounted, as is done initially in Section 2.4.1, the amount of stress taken by the fibres will depend on the fibre lengths.

Any useful theoretical work must include the following:

(1) the prediction of the composite tensile modulus due to long fibre reinforcement;
(2) how this is modified by short fibre reinforcement;
(3) the effect of a spectrum of fibre lengths;
(4) the effect of a spectrum of fibre orientations;
(5) a prediction of tensile strength; and
(6) a prediction of fracture toughness.

These calculations for composite tensile modulus have been made but predictions for tensile strength and fracture toughness are far more difficult even for long fibre reinforcement. In the next sections it will be assumed that the adhesion between the fibres and the matrix is good. The fact that it may not be is an added complication, which makes the composite less strong than it could be.

## 2.4.1. Calculation of Tensile Modulus for Long Fibre Reinforcement

The tensile force $F$ acts on the composite containing long fibres as shown in Fig. 2.3. It is assumed that both the matrix and the fibres are elastic and that their Poisson's ratios are equal. The fibres are all aligned along the direction of the tensile stress. The tensile strain in the composite, that in the matrix and that of the fibres are all assumed equal throughout the material and of value $\varepsilon$. In this situation of equal strain, the total force $F$ is shared between the matrix, $F_m$, and all fibres, $F_f$, giving the tensile force.

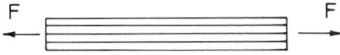

**Fig. 2.3.** A tensile force $F$ acting on a composite consisting of a thermo-plastic matrix surrounding long fibres aligned in the direction of stress

$$F_c = F_m + F_f \tag{2.1}$$

In terms of stresses,

$$\sigma_c A_c = \sigma_m A_m + \sigma_f A_f \tag{2.2}$$

where $A_f$ is the sum of the cross-sectional areas of all the fibres in the composite and $A_m$ is the cross-sectional area of the matrix material. As mentioned earlier,

$$\varepsilon_c = \varepsilon_m = \varepsilon_f = \varepsilon$$

which means that eqn (2.2) can be re-written as:

$$E_c A_c = E_m A_m + E_f A_f \tag{2.3}$$

where $E_c$, $E_m$ and $E_f$ are the tensile moduli of the composite, matrix and fibres respectively.

Rather than working in areas, it is better to use volumes. The total volumes of the composite, $v_c$, of the matrix, $v_m$, and of the fibres, $v_f$, are given by:

$$v_c = l A_c$$

$$v_m = l A_m$$

$$v_f = l A_f$$

If $\phi_m$ and $\phi_f$ are the volume fractions of the matrix and the fibres in the composite, eqn (2.3) can be re-written as

$$E_c v_c = E_m v_m + E_f v_f$$

$$E_c = E_m \phi_m + E_f \phi_f \tag{2.4}$$

This equation represents a simple rule of mixtures, which gives values of $E_c$ for long fibre composites close to values from more rigorous theoretical models and to actual experimental values. More details are given in refs 56 and 57. If the values of the Poisson's ratio, $\gamma$, of the fibres and the matrix are not identical, eqn (2.4) is about 1% in error. For most thermoplastics matrices $\gamma \simeq 0.4$.

In the above it has been assumed that friction at the fibre–matrix interface is sufficient to transfer the load between the matrix and fibres. This is true at low strains. It is found that in glass fibre reinforced polypropylene containing a coupling agent, bonding at the interface increases the composite modulus over a larger temperature and strain range than if no special coupling agent is used.

### 2.4.2. Short Fibre Reinforcement

In the case of continuous fibre reinforcement all the fibres were working at maximum efficiency, with the average strain in the fibre being equal to that in the matrix. In aligned short fibre composites this is not so. Figure 2.4 shows the state of deformation around a short fibre. The fibre restricts the deformation of the surrounding matrix because it is stiffer than the matrix material. The load is transferred from the matrix to the fibre via the interfacial shear stresses. The calculation of the variation of the shear and tensile stresses along a short elastic fibre in an elastic matrix was carried out by Cox,[58] and Fig. 2.5 summarizes the results.

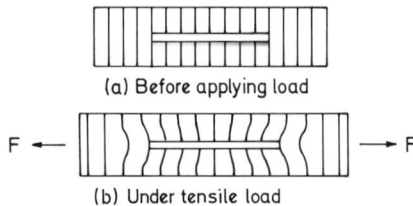

(a) Before applying load

(b) Under tensile load

Fig. 2.4. The effect of a fibre on the matrix deformation.

The shear stress is greatest at the ends of the fibre and decays to zero somewhere along it. The tensile stress is zero at each end of the fibre and reaches a maximum at the centre. If the fibre is just long enough, the maximum tensile stress reaches the tensile stress in the matrix. The ratio $(L/D)_c$ that occurs under these conditions is called the critical elastic aspect ratio. For values of $L/D$ less than $(L/D)_c$ the tensile stress in the

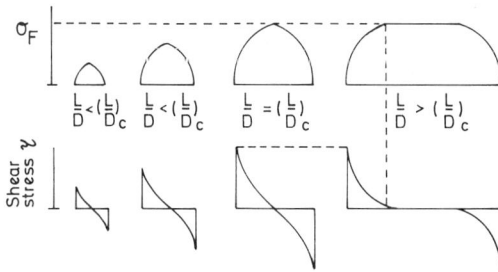

**Fig. 2.5.** The variation of shear stress, $\tau$, and tensile stress, $\sigma_f$, along a short fibre in a matrix for short fibres of varying L/D ratio.

fibre is always less than that in the matrix, and clearly under these conditions the transfer of load from the matrix to the fibre is poor and the mechanical properties of the fibre are not fully utilized. If $L/D > (L/D)_c$, the tensile stress at the interface remains at a maximum over a greater proportion of the fibre length. Here, the transfer of stress from the matrix to the fibre is very efficient but the average tensile stress in the fibre is always less than that in the matrix because of the reduced tensile stress at the ends of the fibre. The efficiency of stress transfer is, therefore, never 100%

In order to accommodate this change in reinforcement efficiency with fibre length, the term $\eta_L$ is included in eqn (2.4). This is the fibre length correction factor and has a value of less than unity.

$$E_c = \eta_L E_f \phi_f + E_m \phi_m \qquad (2.5)$$

The theory developed by Cox[58] gives

$$\eta_L = \left[ \frac{1 - \tanh(\beta L/2)}{(L/2)} \right]$$

where $L$ is the fibre length and

$$\beta = \left[ \frac{2\pi G_m}{E_f A_f \ln(R/r)} \right]^{1/2}$$

where $G_m$ is the shear modulus of the matrix, $r$ is the radius of the fibre and $R$ is the mean separation of the fibres normal to their length.

Figure 2.6 shows the variation of the tensile modulus of the composite, $E_c$, as a function of fibre length, $L$, for aligned glass fibres in polypropylene. In the case of these fibres a length of 2 mm gives the maximum value for $E_c$. It should be remembered that $E_c$ depends on $L/D$ rather than on $L$ alone. This kind of consideration is important when using expensive fibres such as carbon, for if its modulus is not fully utilized, a cheaper substitute would be preferred.

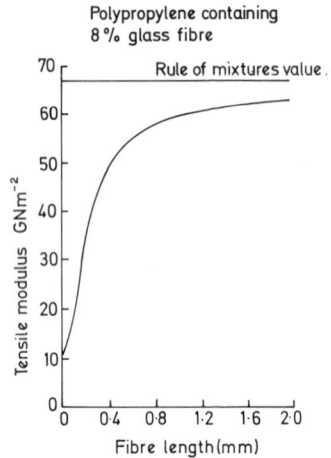

**Fig. 2.6.** The variation of composite tensile modulus, $E_c$, with fibre length, $L$, for glass fibres of fixed diameter, $D$, in a polypropylene matrix.[68]

During processing, such as extrusion or injection moulding, the fibres may be broken, giving smaller $L/D$ ratios than planned. A designer has to be aware of the amount of breakage likely in a given process. In order to model this variation in length the number-average fibre length must be obtained.

Uniaxially aligned composites have highly anisotropic behaviour, with a considerable enhancement of tensile modulus in the direction of orientation (but not in the transverse directions). In these directions the modulus, $E_{cT}$, is dominated by the matrix phase and may be estimated from:

$$\frac{1}{E_{cT}} = \frac{\phi_f}{E_f} + \frac{\phi_m}{E_m} \tag{2.6}$$

In most fractural cases, $E_f \gg E_m$, in which case $E_{cT} \simeq E_m/\phi_m$.

### 2.4.3. Partially Orientated Short Fibres

A further modification to eqn (2.5) is needed to account for the variation in orientation that is a natural consequence of processing:

$$E_c = \eta_0 \eta_L E_f \phi_f + E_m \phi_m$$

where $\eta_0$ is given by.

$$\eta_0 = \sum_{i=1}^{k} a_i \cos^4 \theta_i$$

where $a_i$ is the fraction of the fibres orientated at an angle $\theta_i$ to the direction in which the value of $E_c$ is required. The number of intervals of angle into which the fibre orientation distribution is divided is given by $k$.[59]

Equation (2.6) gives values of $E_c$ very close to measured values when the tensile stress axis coincides with the predominant fibre alignment or for specimens in which the fibre orientation distribution is approximately random-in-the-plane.[60]

A more rigorous treatment of short fibre reinforced composites in which there is a spectrum of orientations and fibre lengths results in the Halpin–Tsai equations and is described in refs 61 and 62. These equations have a form similar to the Krenchel equation and provide a reasonable agreement with experimental values, combined with a simplicity of use. Other more rigorous equations[63, 64] are available but lead to formulae that are more difficult to use. Detailed comparisons between experimental values and theoretical values derived from the various theories are given in refs 60, 65 and 66.

### 2.4.4. Prediction of Tensile Strength

Theories have been developed for the calculation of the tensile strength of short fibre reinforced composites. Consider first the case of continuous fibre reinforcement, as for the calculation of $E_c$ in Section 2.4.1.

Equation (2.2) gives:

$$\sigma_c A_c = \sigma_m A_m + \sigma_f A_f$$

which becomes

$$\sigma_c = \sigma_m \phi_m + \sigma_f \phi_f \tag{2.7}$$

Usually the ultimate tensile strain in the fibre is much less than that of the matrix, and the tensile strength of the fibre, $\phi_{Tf}$, is much greater than that of the matrix, $\sigma_{Tm}$; therefore,

$$\sigma_{Tf} \phi_f \gg \sigma_{Tm} \phi_m$$

and there will be a stress in the matrix $\phi'_m$ at which the ultimate tensile strain is reached in the fibre. From this $\sigma_{Tc}$, the tensile strength of the composite is given by:[67]

$$\sigma_{Tc} = \sigma_m \phi_m + \sigma_{Tf} \phi_f \qquad (2.8)$$

This is a simple mixtures rule similar to the one for $E_c$. When the fibres are not continuous, the average tensile stress on the composite will be given by

$$\sigma_{Tc} = \sigma_m \phi_m + \bar{\sigma}_f \phi_f$$

where $\bar{\sigma}_f$ is the average fibre stress, which is given by:

$$\sigma_f = \frac{1}{L} \int_0^L \sigma_f(x) \, dx$$

If the tensile stress builds up from the fibre ends in the non-linear way shown in Fig. 2.6, then

$$\sigma_f = \sigma_{f\infty} \left[ 1 - (1 - \beta) \frac{L_c}{L} \right] \text{ for } L > L_c$$

where $\sigma_{f\infty}$ is the tensile stress in a continuous fibre in the same matrix under the same loading conditions and $\beta \sigma_{f\infty}$ is the average stress in the discontinuous fibre within a distance $L_c/2$ of either end.

The fibres can be stressed to their tensile strengths when $L > L_c$. If it is assumed that the fibre failure occurs when $\sigma_f = \sigma_{f\infty}$, then substituting in eqn (2.8) gives,

$$\sigma_{Tc} = \sigma_{Tf} \left[ 1 - (1 - \beta) \frac{L_c}{L} \right] \phi_f + \sigma'_m \phi_m \qquad (2.9)$$

A comparison between eqns (2.8) and (2.9) shows that discontinuous fibres provide composites of less strength than continuous ones. However, 95% of the tensile strength of continuous fibre reinforced composites can be obtained when $L/L_c = 10$. In injection moulded products it is not unusual for the process to reduce 80% of the fibres to lengths below $L_c$, so that the strengths of many products are well below the values predicted for continuous fibre reinforcement.[68]

The value of $L_c$ represents the shortest fibre length that may be broken in a given matrix. Fibres below this minimum length are not capable of receiving sufficient tensile stress to break them and composite failure occurs because of failure at the fibre–matrix interface.

Values of $L_c$ are generally larger than predicted values because:

(1) there is debonding at the ends of the fibre where the shear stresses are large;
(2) the interaction between neighbouring fibres constrains the motion of the matrix giving embrittlement; and
(3) the fracture process involves matrix fracture between bundles of fibres rather than just individual fibres.[69]

The tensile strength of a short fibre reinforced thermoplastic decreases as the angle between the fibre axis and direction of loading increases. The tensile strength of the composite in a transverse direction is often less than that of the matrix material above owing to the effect of the fibres.

Prediction of tensile strength as a function of anisotropy is less successful than that of tensile modulus. Further details of the theories involved are given by ref. 69.

## 2.4.5. Fracture Toughness
In many applications the product must be resistant to impact loading and for this an energy absorbing mechanism must be built into the composite. Four possible methods are as follows.

(1) The use of tough matrices, including rubber toughened matrices.
(2) The reduction of the stress concentrating effect by using a soft coating on the fibres that will act as an interlayer.
(3) The use of fibre debonding to absorb impact energy which may cause the fibres to be pulled completely out of the matrix.
(4) The use of a weak interface between the fibre and the matrix such that a crack blunting mechanism takes place.

In the case of (3), the greatest toughness will occur when $L = L_c$ so that the fibre pulls out before it breaks. This will absorb the maximum energy but will not provide as high a tensile modulus as a composite in which the fibres are longer. Clearly, compromise is necessary here.

In the case of (4), the weak interface is not compatible with high stiffness. It would seem that (1) and (2) are the best ways of obtaining improved toughness. In practice the presence of short fibres ($L > L_c$) in products gives rise to enhanced toughness but if the fibre is expensive it is far better to use a shorter fibre of a cheaper material to improve toughness. The use of hybrid composites in which long carbon fibres are combined with shorter, cheaper fibres to give increased toughness has been successful.[70, 71] Richter[71] used intimately mixed carbon and glass fibres in matrices and obtained improved toughness, which was attributed to the higher strain-to-failure of the glass compared with that of the carbon. Multicomponent composites do offer a good opportunity of obtaining an improvement both of modulus strength and of toughness. There are some advantages, however, in reinforcing with flakes or ribbons and these are now discussed.

## 2.5. THEORY OF PLANAR REINFORCEMENT

The main disadvantage of fibre reinforcement is that the mechanical properties are highly directional. In order to achieve multidirectional high strength and high modulus, the planar reinforcement concept was originated. Planar reinforcements include flakes and ribbons.

For planar isotropic composites it might be expected that flakes would be superior to fibres[72, 73] since suitably orientated flakes can stiffen a material in two dimensions rather than in just one. However, some of this advantage is offset because of interactions between nearest neighbours in the three-dimensionally packed structure. This can be explained by considering transfer situations. Riley[9] and Piggot[74] have discussed in detail the stress distributions associated with fibres and flakes respectively.

### 2.5.1. Stress Distribution
Fibres and flakes are similar in the 1–2 plane, but in the 2–3 plane, the flake has two nearest neighbours with which to share stress as compared with the fibre, which has six nearest neighbours (Figure 2.7).

Maximum stress occurs in a reinforcing element where an adjacent element ends, and thus the six nearest neighbours in fibre reinforcement allow a higher average stress in the fibres, whereas the two nearest neighbours in flake reinforcement must take three times the amount of stress transferred compared with the situation in fibre reinforcement.

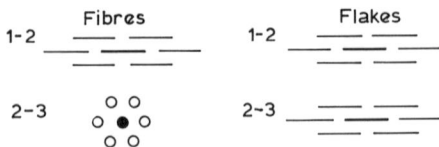

Fig. 2.7. Stress distribution effects in flakes and fibres.

The only difference between fibre and planar reinforcement should be a numerical factor to allow for different geometry and the effect of stress transfer between different planes of flakes. The theory of mechanisms for planar reinforcement is based on fibre reinforcement theory, flake reinforcements being analogous to short fibres, and ribbon reinforcement being analogous to long fibre.

In the following discussion, it is understood that except for the longitudinal ribbon direction, the property relations are similar for flake and ribbon composites.

### 2.5.2. Elastic Properties

A typical arrangement of flakes is seen in Fig. 2.7, and Fig. 2.8 shows the elastic moduli of planar reinforced composites aligned in this manner. Such a composite is anisotropic and requires six moduli to describe its stiffness characteristics:[75]

$E_{11}$, which is Young's modulus parallel to ribbon length; this corresponds to Young's modulus for continuous fibre reinforcement;

$E_{22}$ and $G_{12}$, which are in-plane moduli, and are greatly enhanced for planar reinforcement compared with fibre reinforcement;

$E_{33}$, $G_{13}$ and $G_{23}$, which are the remaining, composite moduli; these values in planar reinforcement are not greatly enhanced over fibre reinforcement.

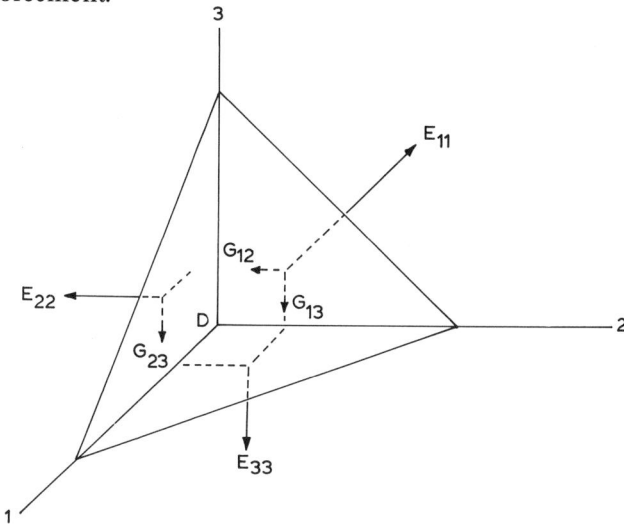

Fig. 2.8. Elastic moduli of planar reinforced composites.

Any useful theory, therefore, must predict the quantities $E_{11}$, $E_{22}$ and $G_{12}$.

In order to calculate the tensile modulus of a composite it is possible to use the 'rule of mixtures'. This ignores any interaction between the constituents of the composites and is normally sufficiently accurate for engineering practice. When applying it to planar composites, however, it is wrong to disregard interactions between the flakes or the ribbons.

It is necessary to modify the rule of mixtures in order to model these systems. This is achieved using a Modulus Reduction Factor [MRF].[75-77] The following equations are derived assuming:

(a) rectangular flakes,
(b) no voids,
(c) no flake imperfections,
(d) ideal flake/matrix adhesion and
(e) uniform flake distribution.

$$\text{MRF} = 1 - \frac{\ln(u + 1)}{u} \tag{2.10}$$

where

$$u = \alpha \left[\frac{G_m \phi_f}{E_f(1 - \phi_f)}\right]^{1/2} \tag{2.11}$$

where $\alpha$ is the aspect ratio ($L/D$ = width/thickness), $G_m$ is the matrix shear modulus, $\phi_f$ is the volume fraction of flake, and $E_f$ is the Young's modulus of the flake.

Introducing the MRF to the rule of mixtures gives:

$$E_c = \phi_f E_f [\text{MRF}] + (1 - \phi_f)E_m \tag{2.12}$$

$E_c$ may be taken in either direction (longitudinal or transverse); thus values of $E_{11}$ or $E_{22}$, respectively, are obtained.

To predict values for $G_{12}$, the Halpin–Tsai equations have been derived,[78] which are based upon the self-consistent micromechanics method developed by Hill.[79]

$$\frac{G_{12}}{G_m} = \frac{1 + \rho \eta \phi_f}{1 - \rho \eta \phi_f} \tag{2.13}$$

where

$$\eta = \frac{G_f/G_m - 1}{G_f/G_m + \rho} \tag{2.14}$$

and $\rho$ is a measure of reinforcement depending upon the boundary conditions; $G_f$ is the flake shear modulus. The appropriate value of $\rho$ required for $G_{12}$ has been shown to be:[78]

$$\log\rho = \sqrt{3}\log\alpha \qquad (2.15)$$

### 2.5.3. Tensile Strength Properties

Although it is possible to apply eqn (2.9) to planar composites, this does not take into account stress concentrations at flake edges, and so the value of transverse tensile strength $\sigma_{22}$ represents an upper bound. As claimed by Riley,[9] the disturbing effects caused by mutual interactions of the flake edges mean that discontinuous reinforced composites can only attain 50–80% of the strength of continuous reinforced composites, depending on the aspect ratio.

For planar composites with a linear elastic matrix, equations have been derived[75, 76, 81] for tensile strength based on the rule of mixtures, assuming:

(a) uniformly spaced flakes parallel to plane,
(b) ideal matrix/flake adhesion,
(c) matrix and flake linearly elastic to failure and
(d) no interactions between reinforcement extremities

and dependent on the failure mode, (i.e. flake fracture or flake pullout), which itself depends on aspect ratio and interfacial shear strength, $\tau$.[80]

$$\frac{W_c}{t} = \frac{\sigma_f}{\tau} \qquad (2.16)$$

where $t$ is the flake thickness, $\sigma_f$ is the rupture strength, and $W_c$ is the critical flake width.

Only if the flake width is greater than $W_c$ will the composite fail by fracture of the reinforcements, otherwise a lower level flake pullout occurs without full utilization of the reinforcement.

### 2.5.4. Flake Fracture

The strength reduction factor accounts for the discontinuous nature of the reinforcement.

$$\sigma_c = \phi_f\sigma_f[\text{SRF}] + \phi_m\sigma'_m \qquad (2.17)$$

$$\text{SRF} = \left[\frac{1 - \dfrac{\tanh(u)}{u}}{1 - \text{sech}(u)}\right] \qquad (2.18)$$

where $u$ is given by eqn (2.11), and $\sigma'_m$ is the tensile strength in the matrix at rupture,

$$\sigma'_m \sim \sigma_m/3$$

($\sigma_m$ is the tensile strength of the matrix).

### 2.5.5. Flake Pullout
Failure of the matrix results in pull out of the flakes and there is little effective strength given to the composite by the reinforcement.

The [MPF] Matrix Performance Factor characterizes this type of failure.

$$\sigma_c = \phi_f \tau_m \, [\text{MPF}] + \phi_m \sigma''_m$$

$$\text{MPF} = \frac{\alpha}{u} \left[ \frac{1}{\tan(u)} - \frac{1}{u} \right]$$

where $\tau_m$ is the shear rupture stress of the matrix or interface, whichever is the lower, and $\sigma''_m$ is the tensile strength in the matrix at rupture.

In this mode of failure $\sigma''_m$ approximates to $\sigma_m$.

### 2.5.6. Toughness
Planar reinforced thermoplastic materials have not been examined in detail for toughness, but soft metal/hard metal laminates have been examined extensively. Figure 2.9 shows the two configurations found to be most effective.

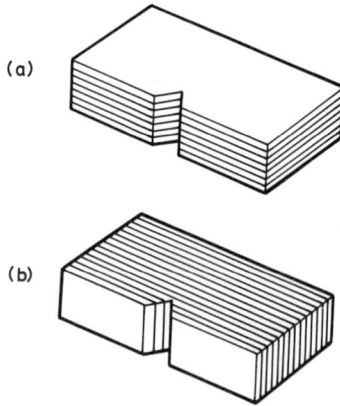

Fig. 2.9. Crack configurations in planar reinforced composites.

Figure 2.9(a) has the crack front perpendicular to the laminae, and crack propagation in the hard laminae is inhibited by the soft laminae. Figure 2.9(b) has the crack front parallel to the laminae; the laminae tend to separate just ahead of the crack, reducing the stress concentration and diverting the crack.

Toughness is considerably improved in both cases. Aligned flakes are subject to similar processes but the exact toughening effect is difficult to quantify.

### 2.5.7. Particle shape

Theory predicts[56, 73] that there will be a difference in strength of a composite reinforced with round platelets as opposed to square platelets, and also a difference in strength of composites reinforced in a regular arrangement compared with irregularly filled composites.

In practice, owing to the inherently irregular flakes associated with the materials involved, these variations are not always apparent.

## 2.6 PLANAR REINFORCEMENTS IN THE FUTURE

Planar reinforcements offer potential advantages in a number of applications, and because planar reinforced composites develop high strength and stiffness omnidirectionally combined with low weight they are potentially most useful for thin-shell flight structures and rotating wings and blades.

Planar reinforcements have the following advantages.

(a) They have higher resistance against liquid or vapour penetration due to the long path-length through the composite.
(b) Higher volume percentages are possible leading to better technical performance.
(c) It is claimed that flakes and ribbons are cheaper to produce than fibres.

Possible choices for ribbons are:

(i) steel ribbons, having one of the highest performances of all reinforcements and also allowing accurate prediction of properties;

(ii)  glassy metals in the form of ribbons, suggesting interesting forms of strong, low-cost ribbon reinforcements,[5] or
(iii) brittle reinforcements such as glass and SiC (but these present the problem of achieving sufficient strength levels, such that overall efficiency will be greater than fibre reinforcement).

What is required to encourage the development of planar reinforcements commercially is a high-performance, low-density ribbon reinforcement analogous to glass fibres, but an immediate short-term possibility would be to use the infinite number of geometric combinations of fibres and ribbons to take advantage of both types of reinforcement.

## ACKNOWLEDGEMENT

Dr H. W. Rayson gratefully thanks Dr H. V. Squires, of CEGB, Scientific Services, Manchester, UK, for the considerable assistance he has given on the survey of carbon fibres.

## REFERENCES

1. Rauch, H. W., Sutton, W. H., McCreight, L. R., *Ceramic Fibers and Fibrous Composite Materials*, Academic Press, New York, 1968, Ch. III.
2. Weeton, J. W., *Machine Design*, 1969 **41**(4), 142.
3. Taylor, G. F., *Phys .Rev.*, 1924, **23**, 655.
4. Evans, C. C. and Parrat, N. J., *New Scientist*, 10 July 1969, 68.
5. Mehan, R. L., *Ceramic Fibers and Fibrous Composite Materials*, Academic Press, New York, 1968, p. 27.
6. Zweben C., *Composite Materials: Testing and Design*, ASTM STP 460, 1969, pp. 528–39.
7. Coleman, B. D., *J. Mech. Phys. Solids*, 1958, **7**, 60.
8. Riley, V. R. and Reddaway, J. L., *J. Mater. Sci.*, 1968, **3**, 41–6.
9. Riley, V. R., *J. Composite Mater.*, 1968, **2**(4), 436.
10. Axelson, J. W., in *Handbook of Fillers and Reinforcements for Plastics*, H. S. Katz and J. V. Milewski (Eds), Van Nostrand, New York, 1978, Ch. 23, p. 415.
11. Axelson, J. W., 'Asbestos in plastics' in *Polymer Plastics Technology and Engineering*, Vol. 4, L. Naturman (Ed.), Marcel Decker and Newall, New York, 1975, p. 93.
12. *A Property Profile of Arpylene Reinforced Thermoplastic Moulding Compounds*, TBA (Turner and Newall) Technical Publication, 1981.
13. Loken, H. Y., 'Asbestos free brakes and dry clutches reinforced with Kevlar aramid fibre', *Earthmoving Industry Conference*, Peoria, IL, April 1980; SAE Technical Paper 800667.

14. Li, L. S., Allard, L. F. and Bigelow, W. C., *J. Macronal Sci. Phys.*, *1983*, **B22**(2), 269–90.
15. Erickson, R. H., *Text. Res. J.*, 1979, **49**(4). Paper presented at the *1st Nat. Fibres Text. Conf.*, Atlanta, Ga, Sept. 1978, pp. 226–32.
16. Dobb, M. G., Johnson, D. J. and Saville, B. P., *J. Polym. Sci., Polym. Phys. Ed.*, 1977, **15**(12), 2201–11.
17. Kulkarni, S. V., Rice, J. S., Rosen, B. W., *Composites*, 1975, **6**(5), 217–25.
18. Greenwood, J. H. and Rose, P. G., *J. Mater. Sci.*, 1974, **9**(11), 1809–14.
19. Hearle, J. W. S. and Wong, B. S., *J. Mater. Sci.*, 1977, **12**(12), 2447–55.
20. Laffitte, M. H. and Bunsells, A. R., *J. Mater. Sci.*, 1982, **17**(8), 2391–7.
21. Bunsell, A. R., *J. Mater. Sci.*, 1975, **10**(8), 1300–8.
22. Neal, T. E., 'Plastics Technology — Recent developments and trends', Tech. Pap. SPE Pac. Tach. Conf., *4th Ann. PACTEC '79*, 1979, pp. 117–20.
23. Erickson, R. H., *Composites*, 1976, **7**(3), 189–94.
24. Takayanagi, M., Kajiyama, T. and Katayose, T., *J. Appl. Polym. Sci.*, 1982, **27**(10), 3903–17.
25. Penn, L., *Adhesia Society, 5th Annual Meeting, Mobile, Ala, USA*, February 1982, pp. 22a–d; [*CONFER*, **9**(12) 4].
26. *Rein. Plast.*, March 1983, **27**(3), 89.
27. Mackay, B. A., *Reinforced Plastics/Composites Institute 37th Annual Conference*, Washington DC, January 1982, Session 24-D, p. 1–5.
28. Lagman, P. L., *Chem. Eng. News*, 1982, **60**(6), 23–4.
29. *Mod. Plast. Int.*, 1983, **13**(9), 7678.
30. Chang, D. C., ASM STP 772, ASTM, Washington, 1982.
31. Gunston, W. T., *Science J.*, 1969, **5**(2), 39–49.
32. Morris, J. B., *Atom*, Oct. 1968, No. 144, 269.
33. Dresher, W. H., *J. Metals*, 1969, **21**(4). 17.
34. Langley, M., *Chartered Mech. Eng.*, 1970, **17**(2), 56–60.
35. Watt, W., Phillips, L. N. and Johnson, W., *The Engineer*, May 1966, **221**(5757), 815.
36. Coyle, R. A. and Gillin, L. M., Department of Supply ARL/MET 63, Australian Defence Scientific Service, Aeronautical Research Laboratories, Feb. 1969.
37. Cooper, G. A. and Mayer, R. M., *J. Mater. Sci.*, 1971, **6**(1), 60.
38. Badami, D. V., *New Scientist*, Feb. 1970, **45**(687), 251.
39. Butler, B. L., *Proc. Conf. Carbon Fibre Technology*, Albuquerque, New Mexico, Jan. 1970.
40. Hawthorne, H. M., Baker, C., Bental, R. H. and Linger, K. R., *Nature*, 1970, **227**(5261), 946.
41. Kawamura, K. and Jenkins, G. M., *J. Mater. Sci.*, 1970, **5**(3), 262.
42. Thorne, D. J., *Nature*, 1970, **225**(5237), 1034.
43. McGowan, H. C. and Milewski, J. V., *24th Annual Technical Conference*, Reinforced Plastics/Composites Division, The Society of the Plastics Industry Inc., Washington, 1969.
44. Williams, W. S., Steffens, D. A. and Bacon, R. *J. Appl. Phys*, 1970, **41**(12), 4893.
45. Jones, B. F. and Duncan, R. G., *J. Mater. Sci.*, 1971, **6**(4), 289.

46. Jones, B. F., *J. Mater. Sci.,* 1971, **6**(9), 1225.
47. de Lamotte, E. and Perry, A. J., *Fibre Science and Technology,* 1970, **3**(2), 157.
48. Moreton, R., *Fibre Science and Technology,* 1969, **4**(1), 1–73.
49. Johnson, D. J. and Tyson, C. N., *Brit. J. Appl. Phys. (J. Phys. D), Ser. 2,* 1969, **2**(6), 787.
50. Hugo, J. A., Phillips, V. A. and Roberts, B. W., *Nature,* 1970, **226**(5241), 144.
51. Wicks, B. J., *J. Mater. Sci.,* 1971, **6**(2), 173.
52. Coyle, R. A., Gillin, L. M. and Bicks, B. J., *Nature,* 1970, **226**(5242), 257.
53. Harling, D. F., *Jeol News,* 1971, **9**(2).
54. Perret, R. and Ruland, W. J., *Appl. Cryst.,* 1969, **2**(5), 209.
55. Tuinstra, F. and Koenig, J. L., *J. Composite Materials,* 1970, **4**, 492.
56. Piggott, M. R., *Load Bearing Fibre Composites,* Pergamon Press, Oxford, 1980.
57. Christensen, R. M., *Mechanics of Composite Materials,* Wiley–Interscience, New York, 1979.
58. Cox, H. L., *Brit. J. Appl. Phys.,* 1952, **3**, 72.
59. Krenchel, H, *Fibre Reinforcement,* Abademisk Ferlag, Copenhagen, 1964.
60. Darlington, M. W., McGinley, P. L. and Smith, G. R., *Plastics and Rubbers Materials Appl.,* 1977, **2**, 51.
61. Kardos, J. L., *CRC Critical Reviews in Solid State Sciences,* 1973, **3**, 419–50.
62. Hopkin, J. C. and Kardos, J. L., *Polym. Eng. Sci.,* 1976, **16**, 344.
63. Chrou, T. W. and Kelly, A., *Ann. Rev. Mater. Sci.,* 1980, **10**, 229.
64. Chow, T. S., *J. Mater. Sci.,* 1980, **15**, 1873.
65. Christie, M. A. and Darlington, M. W., in *Advances in Composite Material,* A. R. Russell *et al.* (Eds), Pergamon Press, Oxford, 1980.
66. Darlington, M. W. and Christie, M. A. to be published.
67. Kelly, A., *Strong Solids,* Oxford University Press, Oxford, 1966.
68. Folkes, M. J., *Short Fibre Reinforcement and Thermoplastics,* Research Studies Press, John Wiley and Sons, New York, 1982.
69. Masoumy, E., Kacir, L. and Kardas, J. L., Dep. Mat. Res. Lab., Washington Univ., Missouri, 1980.
70. Short, D. and Summerscales, J., *Composites,* Jan. 1980, **11**, 33–8.
71. Richter, H., *Kunststoffe,* 1977, **67**, 739–43.
72. Maine, F. W. and Shepherd, P. D., *Composites,* 1974, **5**, 193–200.
73. Glavinchevski, B. G. and Piggot, M., *J. Mater. Sci.,* 1973, **8**, 1373.
74. Piggot, M. R., *J. Mater. Sci.,* 1973, **9**, 1373–82.
75. Rexer, J. and Anderson, E., *Polym. Eng. Sci.,* 1979, **19**, 1–11.
76. Padawar, G. E. and Beecher, N., *Polym. Eng. Sci.,* 1970, **10**, 185–92.
77. Lusis, J., Woodhams, R. T. and Xanthos, M., *Polym. Eng. Sci.,* 1973, **13**, 139–45.
78. Halpin, J. C. and Thomas, R. L., *J. Composite Materials,* 1968, **2**, 488–97.
79. Hill, R., *J. Mech. Phys. Solids,* 1964, **12**, 199.
80. Kelly, A. and Davies, G. J., *Metal. Rev.* 1965, **10**, 1–77.
81. Shepherd, P. D., Golemba, F. J. and Maine, F. W., *Fillers and Reinforcements for Plastics,* Advances in Chemistry Series, 1974.

*Chapter 3*

# Natural Fiber Reinforced Thermoplastics

CARLOS A. CRUZ-RAMOS

*Chemistry Division, Polymers Department, Centro de Investigación Científica de Yucatán, Mexico*

## 3.1. INTRODUCTION

Large amounts of renewable vegetable fibrous matter are produced annually in different areas of the world. For a long time, especially before the advent of synthetic polymers, some of these natural fibers were the main supply for diverse cordage and textile products. In fact, a few of them had found use in ancient times in pottery and structural elements, such as straw clay compositions for bricks,[1] which can be considered as the precursors of modern composite materials.

More recent times have seen the decline in world markets for natural fibers, and there has been a growing interest to find new applications for a certainly abundant and renewable resource. Logically, most of the research in this area has been carried out in those countries where natural hard fibers are available and, even though this research has not attained the degree of sophistication in design and depth already accumulated for advanced composites, the practical implications that stem from understanding and using these materials make it a potentially important field.

Few natural fiber reinforced thermoplastics had been developed until recently. In part, this could be explained in terms of well-known drawbacks of thermoplastics themselves, especially creep,[2,3] which have hindered their more extensive use in composites, when compared with thermosets. Notwithstanding these shortcomings, a renewed interest has developed in thermoplastic composites,[3] thus opening an important field of application for several types of reinforcements, including natural fibers. Indeed, natural hard fibers have been highly regarded as reinforcers of thermosetting matrices[4] and, mainly, for elastomeric ones.[5] Hence, a more pronounced use of these renewable substrates can be expected in the future.

In this review I deal with some of the general characteristics of natural fibers, their physical and chemical features, their potential as reinforcers, and some of their present and foreseeable applications mainly in structural reinforced thermoplastics composites. Selected references to cellulosic pulp fibers extracted from wood have also been included, since their composites are akin to those of hard fibers.

## 3.2. GENERAL CONCEPTS

Natural fibers are biological structures composed mostly of cellulose, hemicelluloses, and lignin. They have been classified from different standpoints.[6,7] One of the most accepted classifications[7] divides natural fibers into three groups: soft seed fibers, soft bast or stem fibers, and hard leaf fibers, according to their origin and properties. Most of the fibers that interest us in this review belong to the latter category. Others, such as flax, cotton, etc., are soft bast fibers used mostly in textile applications.

Among all hard fibers produced, four account for more than 90% of the world production: sisal (*Agave sisalana*), abaca (*Musa testilis*), coconut husk (*Cocus nucifera*) and henequen (*Agave fourcroydes*); the rest is made up by other fibers including ixtle (*Agave lecheguilla*) and jute (*Corchorus capsularis*).[8]

Recent statistics show that 70% of the total world annual production of hard fibers originates in five countries: Brazil, the Philippines, Sri Lanka, Mexico, and Tanzania. On the other hand, most of the total production (90%) is exported to developed countries, among which the United States, France, West Germany, the United Kingdom, Portugal and Japan are the main importers.

Synthetic polymers, mostly those that are fiber-forming, have dominated the world fiber market for the last 40 years, severely affecting the economics of natural hard fibers. Under these conditions, the total production of hard fibers has been monotonically decreasing from a high one million metric tons in the early sixties to a fairly stable 800 000 metric tons in the last few years. Even though they constitute an export commodity, their selling price is low. This can only be supported by the extremely reduced labor costs and the consequent socio-economic implications, which fall beyond the scope of this work.

Efforts have been directed toward carrying out studies on the use of the byproducts in each particular fiber-producing industry and, lately, by trying to find new applications for the fibers themselves, where structural composites might play a relevant role.

## 3.3. CHEMICAL COMPOSITION AND PHYSICAL PROPERTIES

Both soft and hard fibers are built up from varying amounts of cellulose, hemicellulose, and lignin, with minor quantities of other constituents. Table 3.1 illustrates this for several natural fibers.

Each natural fiber is essentially a composite in itself, in which rigid cellulosic ultimate fibers are immersed in a soft lignin matrix. It is

TABLE 3.1
Approximate Chemical Composition of Some Vegetable Fibers (wt %)

| *Fiber* | *Cellulose* | *Hemicellulose* | *Pectins* | *Lignin* | *Water-soluble compounds* | *Fat and waxes* |
|---|---|---|---|---|---|---|
| Sisal[9] | 73·1 | 13·3 | 0·9 | 11·0 | 1·4 | 0·3 |
| Palm[10] | 50–60 | 24 | — | 16–24 | — | — |
| Bagasse[11] | 46 | 30 | — | 15 | 7·0 | 2·0 |
| Jute[9] | 71·5 | 13·4 | 0·2 | 13·1 | 1·2 | 0·6 |

interesting to note that the rigid cellulose backbone, which possesses one of the highest moduli of rigidity of all known polymers,[12] produces fibers with a moderately high strength, owing to the various structural combinations needed to form the microfibrils which are in turn helically wound to form ultimate hollow fibers. The latter are the building blocks of the whole natural fibers that are extracted from the leaves of the producing plant.

Some of the most important physicomechanical properties of the main natural fibers, along with those of wood fibers, are shown in Table 3.2. There are several characteristics worthy of mention from this table: all natural fibers, and wood in particular, have low densities, comparable with the values found only in synthetic polymers. As a consequence, their relative stiffnesses or breaking strengths are high. Also, except for wood, their diameters are relatively large, since most synthetic fibers

TABLE 3.2
Physicomechanical Properties of Some Natural Fibers[9]

| Fiber | Density (g cm$^{-3}$) | Diameter (µm) | Tensile strength[a] (MPa) | Young's modulus (GPa) | Elongation at break (%) |
|---|---|---|---|---|---|
| Sisal (*A. sisalana*) | 1·5 | 10–280 | 507[13] | 16·7[13] | 2–3 |
| Henequen (*Agave fourcroydes*) | 1·49 | 20–370[14] | 580 | 12·8[13] | 3·5–5[14] |
| Coir (husk) (*Cocos nucifera*) | 1·5 | — | 270 | 23·7 | 16 |
| Jute (*C. capsularis*) | — | 30–140 | 900 | 24·1 | 1·5 |
| Ixtle (*Agave lechugilla*) | | | 405[15] | 2·6–5·2[15] | 17–24[15] |
| Bagasse | 1·5 | — | 24[16] | 4·7[16] | — |
| Flax (*Lin usitatissimum*) | 1·52[17] | 40–620 | 900 | 110 | 2–3 |
| Palm (*Yucca carnerosana*) | 1·4[13] | 150–250[13] | 545[13] | 13·7[13] | 3[13] |
| Wood[b] | 0·6[18] | 35[18] | 400[18] | 13·5[18] | — |

[a]Tensile Strength is used here instead of the textile term 'tenacity', for the sake of clarity.
[b]Averages for tropical hardwoods.

designed for use in composites have diameters about ten times smaller than those of vegetable fibers. Aside from indirectly increasing the value of the critical length for reinforcement, another evident problem generated by this occurrence is a more ineffective load-bearing distribution, particularly in short-fiber reinforced materials.

Natural fibers show a high degree of variability in most of their properties. Standard deviations of the order of 50% are not uncommon when determining tenacity or other quantities.[14] As an example, Fig. 3.1 shows the strength and diameter distribution in *A. fourcroydes* class C fibers, two years after being extracted.

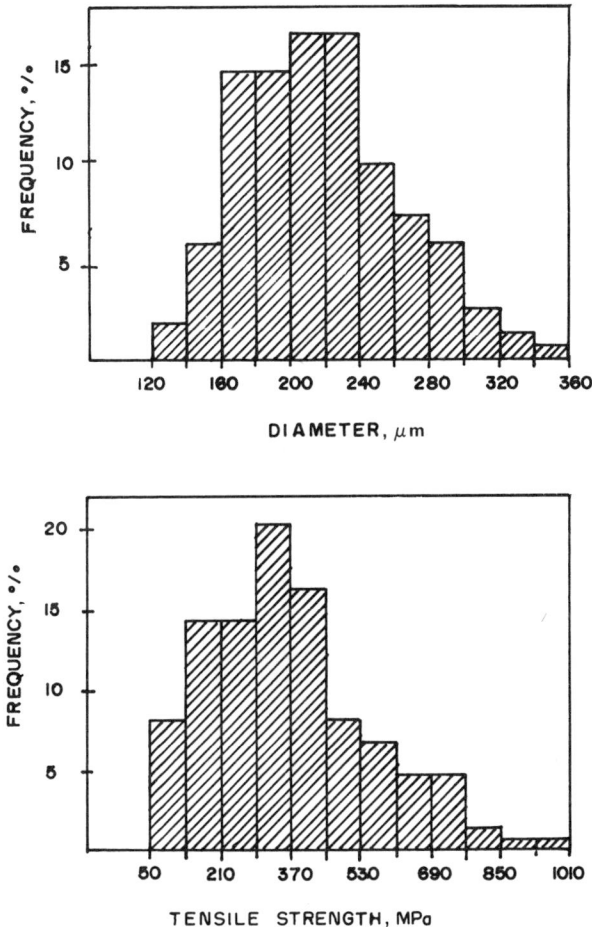

Fig. 3.1. Distribution of properties in class C henequen fibers.

*Carlos A. Cruz-Ramos*

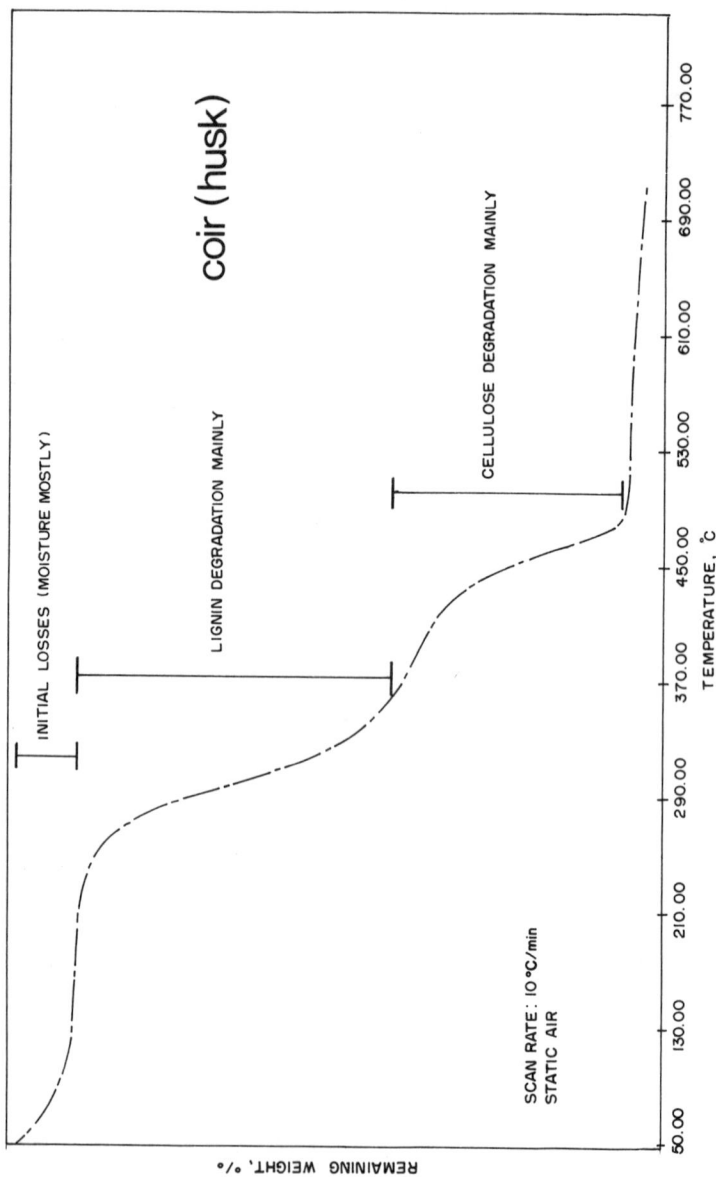

**Fig. 3.2.** Typical thermal degradation pattern for a coir fiber obtained from thermogravimetric analysis.

One further aspect, which is of special relevance to processing, is the high degree of porosity usually found in vegetable fibers in general, which is chiefly a result of the empty lumens of the ultimate cells. Lumen dimensions are variable and depend on the kind of material considered. To cite some examples, palm fibers have a porosity of 10%,[15] similar to that of glass fibers, whereas cotton fibers show values of up to 92%, and still others, such as henequen or sisal, have porosities of the order of 40%.

Thermal properties, in particular thermal or oxidative fiber degradation, have received little attention. However, this is an important design factor which sets the limits of processing temperature and, thereby, the type of matrices that can be reinforced by these fibers. Most natural fibers follow a degradation pattern in air such as that shown for coir in Fig. 3.2. Lignin degradation sets in at around 200°C, and other polysaccharides, mainly cellulose, are oxidized and degraded at higher temperatures.[19] However, most natural fibers lose their strength at about 160°C under these conditions.

Viscoelastic properties have been extensively examined in wood,[20] but not in its single fibers. Nonetheless, it has been pointed out recently that stress relaxation in a natural hard fiber follows a similar pattern to that of a highly crystalline polymer.[21]

On a different approach of fiber utilization, the fiber ultimates that make up the whole fiber can be extracted by means of the traditional chemical pathways used in the paper industry.[22] The fibers obtained in this manner consist mostly of cellulose and are shaped in a ribbon-like form.[10] Their overall characteristics and behavior are similar to those of wood fibers. Apparently, their mechanical properties are superior to those of the whole fiber[12, 21] and, consequently, a higher degree of reinforcement could be expected from them.

With the above elements, one can make a better assessment of the potential of natural fibers for their use in composites.

## 3.4. SOME ELEMENTARY MECHANICS CONSIDERATIONS

Matrix reinforcement from any fiber depends on the relative stiffness and strength of the fiber and the matrix, adhesion between the two, and the aspect ratio of the fiber.[23] To examine appropriately the situation for natural fibers and polymers, it is pertinent to work with a simple idealized model. A suitable situation is that of an elastic matrix with uniaxially-oriented, non-continuous stiff fiber, where the composite strength, $\sigma_c$, can be obtained[23] from the well known expression:

$$\sigma_c = (\phi_f E_f + \phi_m E_m)\varepsilon_c - \frac{\phi_f E_f^2}{4\tau q}\varepsilon_c^2 \qquad (3.1)$$

where subscripts f and m stand for the fiber and the matrix and $\phi$ and $E$ are the respective volume fractions and moduli; $\tau$ is the interfacial fiber–matrix shear stress, and $q$, the fiber aspect ratio (length/radius). $\varepsilon_c$ is the composite unit-elongation.

If large-aspect-ratio ($q \rightarrow \infty$) or continuous fibers are used in the composite, the above equation is reduced to:

$$\sigma_c = (\phi_f E_f + \phi_m E_m)\varepsilon_c \qquad (3.2)$$

which is similar to eqn (2.4).

Hence, to produce reinforcement in this simple case, it is necessary that $E_f > E_m$, as depicted on Fig. 3.3, which reflects the fact that the reinforcing potential of naturally occurring cellulosic fibers is best realized in thermoplastics matrices.[24]

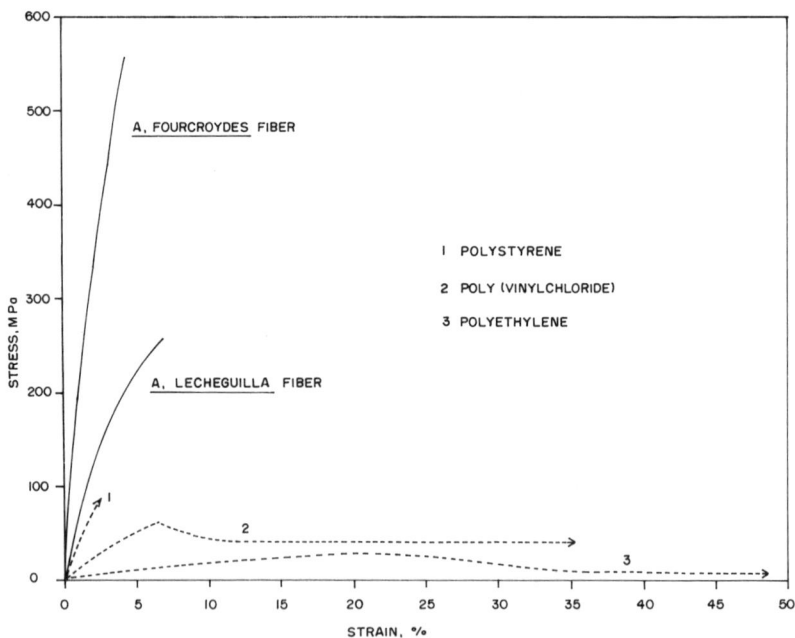

**Fig. 3.3.** Stress–strain behavior of two natural hard fibers in comparison with common thermoplastics.

Another factor that has an important influence on the composite behavior is the interfacial shear stress, $\tau$, which has been experimentally studied in regenerated cellulose fiber systems.[25] The results obtained for the shear stress range from 1 to 8 MPa are in agreement with data for most polymer fiber composites. These values do not help to reduce significantly the negative term in eqn (3.1), and bring about the high dependence of natural fiber composites on the fiber aspect ratio to optimize their mechanical properties.

It has been stated that the minimum relative fiber dimensions required for reinforcement, or critical aspect ratio, $q_{crit}$, can be calculated as:[25]

$$q_{crit} = \frac{\sigma_{fu}}{2\tau} \qquad (3.3)$$

where $\sigma_{fu}$ is the fiber ultimate tensile strength. Hence, for a typical natural fiber with $\sigma_{fu}$ = 500 MPa, and taking $\tau$ as 5 MPa, $q_{crit}$ is nearly 50. Since the diameters for the fibers are in the vicinity of 200 $\mu$m, the net result is a minimum reinforcement length of 5 mm. Still, in order to cancel effectively the adverse adhesion term in eqn (3.1), larger aspect ratios, about 50 $q_{crit}$, are required. These can be easily obtained from hard fibers, but other limitations might be introduced in the composite manufacturing process.

Interestingly, fiber ultimates from hard fibers or wood have definitely higher strengths than those mentioned above, and diameters at least one order of magnitude less, which allow them to be suitable candidates for reinforcement.[12, 21]

## 3.5. MAIN ASSETS OF NATURAL FIBERS

It has been pointed out that cellulosic fibers in general are highly energy efficient from a production viewpoint.[26] Indeed, the amount of fossil fuel needed to produce a natural fiber is only a fraction of what is required to manufacture synthetic fibers, and glass fibers in particular.

It is also evident that the use of vegetable fibers implies the utilization of a renewable resource for both hard fibers and wood fibers alike, with the consequent advantages derived from it. However, hard fibers and wood fibers are in a class by themselves in this regard, since they both come from sources of rapid renewability.[8] Thus, three to seven years are required for a regular hard fiber producing plant to become a supply of fiber material.

More practical advantages can be realized when considering the processing aspects of natural fiber composites: less machine wear than with mineral reinforcements,[27] no health hazards, and a high degree of flexibility. The latter is especially true for the fiber ultimates or cellulosic pulps which will bend rather than fracture during processing, unlike glass fibers.[5] Whole hard fibers suffer from breaking while being intensively mixed with matrix components; nevertheless, this is not as marked as with brittle mineral fibers.[28]

Lighter weight materials are obtained through the use of natural fibers in composites, although with decreased stiffnesses when compared with their glass fiber counterparts.[29, 30]

It is also of interest that the hollow nature of vegetable fibers may impart acoustic insulation properties to certain types of matrices.[31]

Decreased cost has been often mentioned as one of the major reasons why natural fibers should be used in composite materials instead of other fibers.[29, 32] This is true in general, but fluctuating economic conditions make it a somewhat uncertain issue at times.[8]

## 3.6. DRAWBACKS OF VEGETABLE FIBERS

Several disadvantages of natural fiber utilization in composites have been recognized in the past,[1, 4] but a number of solutions to at least control these problems have been proposed more recently and tested in a variety of hard fiber composites.[10, 29, 33]

Hard fiber properties are highly dependent on plant age and changing climatic conditions, which add to the inherent heterogeneities already mentioned.[34] Strict control of cultivating conditions and harvesting is thus important in this regard.[35]

Moisture retention is a major problem in all natural fibers because of decreased wet properties of their composites,[36] and slow hydrolytic degradation of the fibers themselves.[35] Pretreatments with polymers such as polyvinyl acetate or polyvinyl alcohol seem adequately to arrest the problem.[10]

Solar degradation is also an important issue.[32] Ultraviolet radiation is particularly damaging to lignin, which highly absorbs it in the 280–300 $\mu$m region of the spectrum. Mineral loadings that cover the composite have been successfully used to avoid this problem,[37] even though they increase the corresponding composite density.

Biodegradability has long been the major drawback for natural fibers to be more fully utilized in composites.[1] Biodeterioration of hard fibers apparently follows first-order kinetics,[29] but is actually a complex process in which enzyme inhibition,[38] lignin, and other factors play a role,[39] and only general solutions have been attempted with moderate success for this problem. The addition of zinc chloride in relatively small amounts hinders biodegradation to some extent;[10, 29] glass fiber gel coating,[40] used to increase weathering resistance in other composites, also seems to be a suitable solution, since biodeterioration is connected to some extent with the presence of water. Crystalline thermoplastics should be effective in this regard because of their decreased permeation. Also, fluoropolymer and acrylic films have been suggested as possible barriers,[29] but have not been tested.

## 3.7. PROCESSING

### 3.7.1. Types of Fiber and Preforms for Composites
Natural fibers can be made available in different forms to produce composites. Long[15] or short fibers,[11] felts or paddings,[32] extracted ultimates, or pulps, either as sheets[33] or as separate fibers,[27] have been used.

A distinction must be made at this point between whole natural hard fibers, whose diameters are of the order of 200 $\mu$m, and fiber ultimates, with diameters around 20 $\mu$m.

Other types of shapes, such as crêped sheets from ultimates[41] and thin whole-fiber mats, that preserve high aspect ratios and are used in thermosetting compositions, are also good candidates for thermoplastics.[13]

Continuous single-filament winding operations for composite formation are not feasible, since the maximum length that a whole natural hard fiber can achieve is about 2 m.[35] It should also be stressed that hard fibers are mechanically weakened by an increased number of processing steps; hence, it should always be kept in mind that, in general, preforms contain fibers of reduced relative strength.

### 3.7.2. Fiber Pretreatment
One important difficulty that has prevented a more extended utilization of natural fibers is their lack of good adhesion to most matrices.[42] A feasible, albeit expensive, method of enhancing the matrix–fiber

interaction is pregrafting of the fiber with the polymer to be used in the composite.[27, 43]

Other approaches have been focused on the presence of the hydroxyl groups of cellulose. For example, the use of reactive functional oligomers or monomers of the isocyanate type has been reported as an aid in fiber-to-matrix adhesion,[5, 44] as well as polyesteramide polyols.[42]

The addition of small amounts of low molecular weight polyols has also been tested with apparent success.[45] Still other pretreatments involve the application of polymeric coatings.[10] In general, this should be a fertile area for further research.

### 3.7.3. Fabrication Methods

Melt processing has been used in two distinct modalities to produce natural fiber thermoplastic composites: hot-press lamination and extrusion.

Polyethylene- and polyvinyl chloride–natural fiber– sand laminates have been prepared by adding the polymeric powders into felts, followed by high-temperature pressing in a mold.[32, 46] Also, intensive mixing of short ultimate fibers and the polymeric powders has been performed prior to hot pressing, in order to obtain a more homogeneous material.[27]

Extrusion of short cellulosic fibers and polyvinyl chloride has been tested as a means to produce materials with a highly oriented reinforcement, through the use of a special die.[5] This process is particularly amenable to the use of short fiber ultimates or even short hard fibers, since they retain a relatively high degree of their original strength and length in contrast to more rigid reinforcements.[5, 47]

Fiber- or particle-boards belong to a distinct, although parallel, area of composites. Nevertheless, it is well known that improved materials can be obtained if the already-formed board, or wood, is impregnated with considerable amounts of vinyl-type monomers that are then polymerized to form dimensionally stable compositions, which are true hybrids between typical fiber-boards and reinforced thermoplastics. Through the literature these materials are known as wood–plastic combinations,[48] or wood–plastic composites (WPC). Reference is made to their processing here, because their preparation processes might become applicable to hard fibers in the future.[49]

*In situ* polymerization of these materials involves γ-irradiation[50] and chemically activated initiation.[51] An excellent monograph on the topic, on which numerous references appear in the literature, was edited by the IAEA,[52] and a more updated review was recently published.[48] Even though *in situ* polymerization is an obvious route to circumvent the difficult penetrativity of high polymers into fibrous substrates, care must be taken to follow certain conditions, such as removing all possible radical scavengers, drying the fibers, and avoiding the presence of air or oxygen, which adversely affect the polymerization process.[52]

## 3.8. MATERIALS PRODUCED, PROPERTIES AND APPLICATIONS

Most of the natural fiber reinforced materials mentioned in the literature belong to the thermosetting matrix category;[10, 29, 53-56] however, there is an increasing number of recent reports regarding thermoplastic-based composites with natural fibers. Poly(styrene-*co*-acrylonitrile) (SAN) and ABS resins have been evaluated as binders for bagasse fiber and other agricultural residues.[57, 58]

Polyethylene–henequen–sand laminates appear to have a relatively low density, high weathering resistance and low water absorption.[32] Ixtle–polyethylene, coir–polyethylene, and henequen–polyvinyl chloride combinations are under study by the same group of researchers.[37] PVC–ixtle–sand laminates[46] show good mechanical properties and are apparently suitable for use as construction materials. In general, building, and especially roofing, applications are expected for these composites.

Using pine kraft lap pulp sheets as a basis, polyethylene laminates which show good flexural properties at low humidities have been produced for application in building and packaging.[36] The decline in the mechanical properties for these materials at high degrees of humidity has been partly eliminated by pre-acetylation or grafting of the fibers.[33]

Pregrafted and ungrafted pulp fibers from aspen and spruce have been combined with polystyrene by hot pressing[27] and, as a result, a general improvement of mechanical properties was observed at high fiber contents. The future applications for these materials range from storage bins to car components.[27]

Extruded composites of plasticized polyvinyl chloride and short cellulose fibers have been investigated by Goettler.[5] Pronounced increases in tensile modulus and both yield and ultimate strength are observed. Single-step processing of reinforcement and polymer, with good product performance, are the key characteristics of the material, whose field of application lies in the vinyl hose industry.[5]

Table 3.3 gives an overview of the mechanical properties of some of the above composites.

TABLE 3.3

Mechanical Properties of Selected Natural Fiber–Thermoplastic Combinations[a]

| Composite | Approximate composition (wt %) | | | Mechanical properties | |
| --- | --- | --- | --- | --- | --- |
| | Fiber | Polymer | Filler | | |
| Ixtle–polyvinyl chloride–sand[46] | 10 | 40 | 50 | Flexural strength, MPa | 32 |
| | | | | Elastic modulus, MPa | 1430 |
| | | | | Shear strength, MPa | 2·0 |
| Aspen pulp–polystyrene[27] | 40 | 60 | — | Tensile strength, MPa | 45 |
| | | | | Tensile modulus, MPa | 55 |
| | | | | Energy at break, kJ/m$^2$ | 15 |
| Santoweb-W (treated cellulose)[5] | 21 | 79[a] | — | Tensile strength, MPa | 19·0 |
| | | | | Tensile modulus, MPa | 131 |
| | | | | Ultimate elongation, % | 18·2 |

[a]Preformulated, plasticized polymer with added stabilizers and colorants; fibers aligned in parallel.

Wood–plastic combinations involving oak, pine, poplar, bamboo, and other species have been prepared[52, 59, 60] with different contents of polymer. Styrene, acrylonitrile, methyl methacrylate, vinyl acetate, and styrene–acrylonitrile mixtures appear to be the preferred monomers,[52] with the general result of an improvement of mechanical properties.[61] However, the inclusion of macromers such as polyoxyethyleneglycol methacrylate (PEGMA) to enhance dimensional stability further has been considered more recently.[20]

### 3.9. FINAL REMARKS

Natural fibers possess a variety of appealing properties to be used in thermoplastic composites of at least moderate strength. The handful of examples of cellulosic fiber–polymer combinations reported, most of which are still in the process of being developed, show that the interest in using them in composites has been growing in recent times. In the end, the renewability of natural fibers combined with the reprocessability of thermoplastics might play the key roles behind further progress in the field.

## REFERENCES

1. Gordon, J. E., *The New Science of Strong Materials*, Walker, New York, 1968, pp. 167, 171.
2. *The Composite* published by McLean Anderson, 1984, **3**(4), 1.
3. Muzzy, J. D. and Kays, A. O., *Polymer Composites*, 1984, **5**, 169.
4. Lubin, G., *Handbook of Fiberglass and Advanced Plastics Composites*, Robert E. Krieger, Melbourne, Florida, 1969, pp. 42, 150, 370.
5. Goettler, L. A., *Polymer Composites*, 1983, **4**, 249.
6. Kirby, R. H., *Vegetable Fibers*, Leonard Hill, New York, 1963.
7. Dewey, L. H., El henequen, *Las Principales Fibras Vegetales*, 1916, **1**(8).
8. Garcia de Fuentes, A. and De Sicilia, A., *El Mercado Mundial de Fibras Duras*, Centro de Investigación Científica de Yucatán, A. C. Mérida, Yucatán, Mexico, 1984.
9. *Kirk–Othmer Encyclopedia of Chemical Technology*, 3rd edn, Vol. 10, 1980, p. 182.
10. Belmares, H., Barrera, A. and Monjaras, M., *Ind. Eng. Chem. Prod. Res. Dev.*, 1983, **22**, 643.
11. Salyer, I. O. and Usmani, A. M., *Ind. Eng. Chem. Prod. Res. Dev.*, 1982, **21**, 13.
12. Page, D. H., El-Hosseiny, F., Winkler, K. and Lancaster, A. P. S., *Tappi*, 1977, **60**, 114.
13. Belmares, H., Barrera, A., Castillo, E., Verheugen, E., Monjaras, M., Patfoort, G. A. and Bucquoye, M. E. N., *Ind. Eng. Chem. Prod. Res. Dev.*, 1981, **20**, 555.
14. Cruz-Ramos, C. A., Moreno, E. and Castro, E., *Memorias del Primer Simposio Nacional de Polimeros*, UNAM, Mexico, D. F., 1982, p. 153.
15. Belmares, H., Castillo, J. E. and Barrera, A., *Textile Res. J.*, 1979, **49**, 619.
16. Valadez, C. E., M. Sc. Thesis, Facultad de Ciencias, Universidad Nacional Autónoma de México, 1981, p. 59.
17. Sadov, F., Korchagin, M. and Matetsky, A., *Chemical Technology of Fibrous Materials*, Mir Publishers, Moscow, 1978, p. 122.
18. Tamolang, F. N., Wangaard, F. F. and Kellogg, R. M., *Tappi*, 1967, **50**(2), 68.
19. Akita, K. and Kase, M. J., *Polym. Sci., A-1*, 1967, **5**, 833.
20. Handa, T., Yoshizawa, S., Seo, I. and Hashizume, Y., *Org. Coat., Plast. Chem. Prep.*, 1981, **45**, 375.
21. Cruz-Ramos, C. A., In: *Advances in Rheology IX*, Vol. 3, eds B. Mena, A. García-Rejón and C. Rangel-Nafaile, Elsevier, New York, 1985, pp. 733–40.
22. Casey, J. P. (Ed.), *Pulp and Paper Chemistry and Technology*, 2nd edn, Vol. II, Wiley–Interscience, New York, 1980.

23. Piggott, M. R., *Load Bearing Fibrous Composites,* Pergamon, Toronto, 1981, p. 168.
24. Goettler, L. A. and Shen, K. S., *Rubber Chem. Technol.,* 1983, **56**, 619.
25. Westerlind, B., Rigdahl, M., Hollmark, H. and De Ruvo, A., *J. Appl. Polym. Sci.,* 1984, **29**, 175.
26. Belmares, H., in *Biotecnología y Aprovechamiento Integral del Henequén y otros Agaves,* C. Cruz, L. del Castillo, R. Ondarza and M. Robert (Eds), Centro de Investigación Científica de Yucatán, A. C. Mérida Yucatán, Mexico, 1985.
27. Kotka, B. V., Che, R., Deneault, C. and Valade, J. L., *Polymer Composites,* 1983, **4**, 229.
28. Rao, R. M. V. G. K., Balasubramanian, N. and Chanda, M., *J. Appl. Polym. Sci.,* 1981, **26**, 4069.
29. Belmares, H., Barrera, A. and Monjaras, M., *Ind. Eng. Chem. Prod. Res. Dev.,* 1983, **22**, 652.
30. Persson, H. and Skarendahl, A., *UNIDO International Forum on Appropriate Industrial Technology,* ID/WD 282/11, UNIDO, New Delhi, 1978.
31. FAO, *Research Series on Hard Fibers,* No. 7, WM/A4939/c, 1970.
32. Padilla, A. and Sanchez, A., *J. Appl. Polym. Sci.,* 1984, **29**, 2405.
33. Michell, A. J., Vaughan, J. E. and Willis, D., *J. Appl. Polym. Sci.,* 1978, **22**, 2047.
34. Crane, J. C. and Wellman, F. L., *Turrialba,* 1950, **1**(2), 74.
35. Lock, G. W., *Sisal,* Longmans, Green and Co., London, 1969.
36. Michell, A. J., Vaughan, J. E. and Willis, D., *J. Polym. Sci., Polym. Symp.,* 1976, **55**, 143.
37. Padilla, A. and Fuentes, P., in *Simposio International sobre la Biotecnología y Perspectivas Tecnológicas del Henequén y otros Agaves,* C. Cruz, L. del Castillo, R. Ondarza and M. Robert (Eds), Centro de Investigación Científica de Yucatán, Mexico, 1985.
38. Focher, B., Marzetti, A., Cattaneo, M., Beltrame, P. L. and Carniti, P., *J. Appl. Polym. Sci.,* 1981, **26**, 1989.
39. Buchholz, K., Puls, J., Godelmann, B. and Dietrichs, H. H., *Process Biochemistry,* Dec/Jan. 1980/81, p. 37.
40. Blaga, A. and Yamasaki, R. S., *Mater. Constr. (Paris),* 1978, **11**(63), 175.
41. Hilton, R. D., US Department of Commerce NTIS PB-226 967, 1971.
42. Mukherjea, R. N., Pal, S. K. and Sanyal, S. K., *J. Appl. Polym. Sci.,* 1983, **28**, 3029.
43. Gaylord, N. G., US Patent 3645939, 1972.
44. Hse, C.-Y., US Patent 4209433, 1980.
45. Padilla, A., Fuentes, P., Sanchez, A. and Cardoso, J., *Proc. First National Polymer Meeting — Mexico,* Universidad Autonoma Metropolitana — Iztapalapa, 1982, p. 140.
46. Padilla, A., Sanchez, A. and Castro, M., In: *Advances in Rheology IX,* Vol. 3, eds B. Mena, A. García-Rejón and C. Rangel-Nafaile, Elsevier, New York, 1985, p. 749.
47. Chakraborty, S. K., Setua, D. K. and De, S. K., *Rubber Chem. Technol.,* 1982, **55**, 1286.
48. Schaudy, R. and Proksch, E., *Ind. Eng. Chem. Prod. Res. Dev.,* 1982, **21**, 369.

49. Sadurní Castillo, P. and Acevedo Reyes, D., *Rev. Soc. Mex. Quim.*, 1979, 23(2), 88.
50. Wang, U.-P., in *Impregnated Fibrous Materials*, The International Atomic Energy Agency, Vienna, 1968, p. 35.
51. Stannett, V. T., in *Impregnated Fibrous Materials*, The International Atomic Energy Agency, Vienna, 1968, p. 45.
52. International Atomic Energy Agency, *Impregnated Fibrous Materials*, Proceedings of a Study Group, IAEA, Vienna, 1968.
53. Hamed, P. and Coran, A. Y., in *Additives for Plastics*, R. B. Seymour (Ed.), Academic Press, New York, 1978, p. 29.
54. Semsarzadeh, M. A., Loftali, A. R. and Mirzadeh, H., *Polymer Composites*, 1984, 5, 141.
55. Murty, V. M. and De, S. K., *Rubber Chem. Technol.*, 1982, 55, 287.
56. Coran, A. Y., Boustany, K. and Hamed, P., *Rubber Chem. Technol.*, 1974, 47, 396.
57. Usmani, A. M., Ball, G. L. III, Salyer, I. O., Werkmeister, D. W. and Bryant, B. S., *J. Elastomers Plast.*, 1980, 12, 18.
58. Usmani, A. M. and Salyer, I. O., *Prepr. Amer. Chem. Soc. Div. Org. Coat., Plast. Chem.*, 1981, 45, 459.
59. Boyle, W. R., Winston, A. W. and Loos, W. E., Report (ORO-2945-10), Contract AT(40-1)-2945 for the US Atomic Energy Commission, West Virginia University, 1971.
60. Hills, P. R., Barrett, R. L. and Pateman, R. J., Report AERE-R-6090, UK Atomic Energy Agency, 1969. (Available through US National Technical Information Service.)
61. Manson, J. A. and Sperling, L. H., *Polymer Blends and Composites*, Plenum, New York, 1976, p. 343.

*Chapter 4*

# Continuous Fibre Reinforced Thermoplastics

F. N. COGSWELL

*Imperial Chemical Industries PLC, New Science Group, Wilton, UK*

## 4.1. INTRODUCTION

In continuous fibre reinforced plastics it is the fibre which determines the basic property profile: the function of the resin is to allow the fibre to develop its full potential by transferring the load from one fibre to another and, by providing a support, to prevent the fibre from buckling. From this standpoint it is clear that the best possible property profile will derive from the highest loading of the longest fibre, provided that each fibre is fully wetted by a resin of adequate stiffness to prevent buckling of the fibres when subjected to compressive loading. Further, we require that the resin shall maintain the ability to continue to transfer load from one fibre to its neighbours in a service environment.

Besides its function as a glue, the resin phase also acts as a carrier for the fibre at the fabrication stage: it is responsible for seeing that the fibre goes where it is directed, and stays there. The resin phase, and, in particular, how it is made mobile to shape the product and how it is made form-stable to retain that shape, also determines the economics of fabrication.

A resin system to work with high loadings of continuous fibre reinforcement may be chosen from three groups:

(1) 'chemi-setting' systems based on the mixing of two components which react and polymerise;

(2) 'thermo-setting' systems which polymerise when heated, usually forming a crosslinked product; and

(3) 'thermoplastic' resins derived from the rigorous chemistry of linear chain polymerisation producing a system which softens when heated in a physically reversible process.

On a snapshot impression, the product of the fibre reinforced plastics industry can be divided into two groups:

(1) low loadings (< 30% by volume) of short (< 1 mm) fibres based on thermoplastic resins and injection moulding technology, or using chemi-setting systems with the emerging reinforced reaction injection moulding (RRIM) technology; and

(2) high loadings (> 50% by volume) of long (including continuous) fibre based on thermosetting resins.

The preference for thermoplastic or chemi-setting resins for work with low loadings of short fibre is a tacit recognition of the cycle time advantages in fabrication for a system which requires only heat exchange or only chemical reaction to effect the transition from a mobile to a solid state in comparison with a system which requires heat exchange in order to initiate a chemical reaction. The historical preference for thermosetting resins with high loadings of long fibres is an acknowledgement of the convenience of impregnating the fibrous matt with a stable pourable resin, where surface tension can be relied upon to effect the wetting, compared with the difficulty of impregnating a closely packed array of fibres with a viscous resin or with one which is reacting as it is flowing into the spaces between the fibres. In respect of the latter option, impregnation of fibres by a chemically reacting mixture, the acrylic resins of the MDR series[1] appear to be the newest development. Given the relative newness of RIM technology, the impregnation of fibres by a chemically reacting mixture may, perhaps, form another chapter of another book. By contrast the possibilities of continuous fibre reinforced thermoplastics have, for many years, attracted numerous workers.[2-5]

Thermosetting resins, and in particular epoxies, during the last decade have provided the major source of matrix materials for high-performance structural composites based on continuous fibre reinforcement. Those resins have demonstrated a good balance of properties combined with convenient processing technology: the high performance of components made from such resins has generated a new design philosophy and a potential demand for an appropriate mass production process. As with any new high-performance material the design philosophy has pushed the material to its limits and identified certain property constraints — shelf life, water sensitivity and brittleness — which appear to be inherent in that family. Looking forward to the next generation of materials, thermoplastic resins offer an attractive balance of properties.

The advantages offered by thermoplastic resins fall into four groups. The rigorous chemistry of linear chain polymerisation gives composite materials indefinite shelf life from which the user obtains enhanced quality assurance. Toughness of the resin phase can translate into improved damage resistance in the product, and crystallinity of the resin phase gives potential for excellent environmental resistance. Damage tolerance and environmental resistance allow the designer increased freedom to extend the limits of performance. Finally, the thermoplastic characteristic of the resin permits rapid processing, the ability to reclaim scrap via injection moulding, and the opportunity to repair damaged structures. Taken together these translate into economic and versatile fabrication technologies. The one readily identifiable disadvantage of thermoplastic systems is the need to work at high temperatures (as high as 380°C with polyether ether ketones), but while this removes the convenience of hand lay-up the potential for automated tape laying and filament winding technologies is not affected. These advantages may be viewed as decisive factors in determining the future growth of the fibre reinforced plastics industry.

The major developments in fibre reinforced plastics will always be associated with developments of improved fibres, since it is the fibres which dominate the mechanical property profile. The last decade has seen significant advances in the technology of producing reinforcing fibres, especially the production of fibres of higher strength. The provision of better fibres does not reduce the stress on the matrix phase. The greater the inherent property of the fibre, the greater is the demand on the matrix to resist damage from impact or from hostile environments and to transmit higher stresses.

Although we can detect a strong potential need for thermoplastic resin composites there is one problem to be overcome: how to impregnate the fibrous bed with a viscous resin. The issue of viscosity is an anomaly in the whole area of continuous fibre reinforced plastics. The anomaly is apparent when we see that for conventional thermo-setting systems the low viscosity state, which permits impregnation to be achieved rapidly, is actually an embarrassment at the later fabrication stages if it allows the fibres to move from their nominated post or the resin to be squeezed back out of the fibres. Not surprisingly, some of the most recent developments in thermosetting resins appear to be directed at increasing the viscosity of the resin to give a more tractable prepreg. Thus, low viscosity is required at one stage only — the formation of a 'prepreg' — and that stage is not of interest to the user of the material. This is not to minimise the difficulty of the viscosity problem. In the case of impregnation with thermosetting resins a viscosity of below $0.1$ Ns/m$^2$ appears to be preferred and the process is frequently assisted with a solvent. Most commercial thermoplastics have viscosities in the range 100 to 100 000 Ns/m$^2$. We may thus define the 'viscosity' problem as representing a difficulty factor between one thousand and one million.

The 'viscosity' problem can be phrased in more homely terms. Suppose we wish to make a piece of prepreg to the size of a sheet of A4 typing paper of good quality so that it is about $0.3$ m long, $0.2$ m across and $0.1$ mm thick. This sheet of prepreg is to contain 50% by volume of $7 \, \mu$m diameter fibres (just visible to the naked eye). We can then calculate that we have about three millilitres (a small thimbleful) of resin to spread over a total surface area of about two square metres (a reasonable size dining table). For the traditional thermosetting technology our resin has the consistency of a rather sugary liqueur and we can smear it over the whole table (at a thickness of $2.5 \, \mu$m) and still spill some on the carpet. For a thermoplastic material we must replace the liqueur with an equal volume of chewing gum — the first problem is getting it out of the thimble. Thus we have the challenging situation of an urgent need combined with a genuinely sticky problem.

## 4.2. A SURVEY OF IMPREGNATION STRATEGIES

The definition of a 'viscosity' problem identifies one immediate solution — if you push hard enough for long enough, then to quote one of the oldest philosophies 'everything flows': note, however, that Heraclitus is, perhaps, more correctly translated as 'everything is in a state of flux', referring thereby to technological and social change rather than a specific rheological phenomenon. This approach is the basis of the great majority of current published work on continuous fibre reinforced thermoplastic composite made by the technology of 'film stacking'.[2-6] In 'film stacking', layers of fibre are interleaved with layers of thermoplastic film and subsequently compression moulded at pressures usually reported to be about $10^7$ N/m$^2$ (100 atmospheres) for times of about one hour.

Early development of this tactic was pioneered at Rolls Royce[6] and at the Royal Aircraft Establishment[2] and the technology was successfully commercialised by Specmat Ltd[5] in the United Kingdom. Compression moulding is a batch, rather than a continuous, process, the size of the 'batch' being determined by the capacity of the press: for processes working at pressure of 100 atmospheres each square metre requires a locking force of 1000 tonnes. A continuous product could be obtained from a semicontinuous process, stepping the product through a press so that each moulding overlapped the one before it. Smith[7] suggests that the principles of film stacking can be transcribed to a genuinely continuous lamination process.

The 'viscosity' problem can be reduced by the use of solution technology analogous to that employed with thermosetting resins. That technology was pioneered by Turton and McAinsh[8] to make thermoplastic composites based on aromatic polymers such as polyether sulphone. The use of solution technology implies a limitation to making thin 'prepreg' from which the solvent can be extracted, itself a difficult operation,[9] and the convenience of impregnation at low temperatures with dissolved resins carries with it the potential embarrassment that, unless the resin is subsequently crosslinked, the product may be subject to environmental attack by a similar solvent. Thus the technology is constrained to making products for applications where resistance to common solvents is not a major concern, or alternatively it requires the selection of a special solvent to dissolve a resin which has strong resistance to common solvents.

With a significant part of the thermoplastics industry based on electrical insulation of wire it is not surprising to find that coating of fibre bundles has been a frequent approach to the production of fibre reinforced thermoplastics.[10-12] Whilst it is easy to obtain bundle impregnation by this route, a common experience is that full impregnation of the ten thousand fibres in the tow usually requires a subsequent high pressure moulding operation analogous to 'film stacking' to obtain the best possible results. The idea of coating individual fibres in a cross head, 'wire-covering', process and subsequently combining these into a sheet is certainly an option available with large diameter ($> 100 \, \mu$m) monofilament but is impracticable for conventional fibres ($\sim 7 \, \mu$m diameter).

Of process technologies which accept the intractability of the viscosity problem the impregnation of the fibres by fine particles has many champions. The fibre bundle can be impregnated by immersing it in a suspension of particles supported by an inviscid medium including such options as a fluid bed[13,14] or water slurry.[15] If the powder particles can be made sufficiently fine ($0 \cdot 5$–$2 \, \mu$m) then it is, in principle, possible to obtain a totally wetted product after the sintering stage. Fine nylon powders designed for this process are now commercially available.[14]

The synthetic fibre industry provides an alternative strategy for impregnation based on mixing fibres of the thermoplastic with those of the reinforcement. A fabric woven from such a mixed fibre product does undoubtably have interesting handling properties; however, the statistics of fibre mixing make the construction of the mixed fibre tow a demanding technology if a good quality of fibre wetting is to be achieved in the final sintered product. The alternative of coweaving tows of reinforcement and resin fibres requires subsequent moulding cycles of the same order as are used with 'film stacking', the main advantage being excellent drapability in a well designed weave.

A final approach to the production of continuous fibre reinforced thermoplastics involves impregnation by a prepolymer followed by condensation to give a fully polymerised linear chain thermoplastic matrix.

With such a diversity of possible approaches any listing of technologies cannot be exhaustive. In developing an optimum technology it is essential to keep the product target firmly in sight. We require a product based on about 50% by volume of continuous fibres wherein each fibre is fully wetted by a thermoplastic resin able to withstand the rigours of a designated service environment.

## 4.3. THE PRESENT STATE OF THE ART

Thermoplastic composites based on long fibre reinforcement are firmly established as stampable sheet products including 'Azdel'[16] and 'STX'.[17] These products contain a proportion of truly continuous fibre blended with fibre of intermediate length in knitted or random matt form. Containing about 25% by volume of fibre, they form an intermediate group between low loadings of short fibres and high loadings of continuous fibres.

One constraint of a high loading (50% by volume) of continuous fibre is that, for space filling reasons, they require a high degree of fibre collimation followed by weaving or laying up to produce a more isotropic structure. For several years the only commercially available product in this field has been a sheet stock material fabricated by Specmat Ltd using film stacking technology.[5] In 1982 ICI introduced 'Aromatic Polymer Composite APC-1',[18] based on 52% by volume of continuous carbon fibre reinforcement in 'Victrex' polyether ether ketone, a thermoplastic composite for aerospace applications: this product is available as a preimpregnated tape wherein all the fibres are effectively wetted by the resin so that lamination requires only nominal pressures. APC-1 is also available as laminated sheet stock. Also in 1982, Polymer Composites Incorporated[12] introduced a range of continuous fibre reinforced thermoplastics, 'Fiberod', claimed to be 'the ultimate in thermoplastic composites'. This product is apparently produced by a melt coating process and requires a further high-pressure consolidation process to develop 'the full potential' of the fibre reinforcement.[19] A range of fibre reinforced thermoplastic tape products under the name 'SPIFLEX' were introduced by SPIE–Batignolles in 1983:[21] these tapes, which contain up to 50% by volume of fibre reinforcement, are mainly designed for overwrapping of pipes. The year 1983 saw the launch by Phillips[20] of a range of thermoplastic composites based on 'Ryton' polyphenylene sulphide including high loadings of continuous carbon fibre. Research activity in this field is intensifying, including a cooperative programme at Battelle Columbus laboratories and Defense Agency sponsored programmes in the United States.[22]

To illustrate the property profile available to continuous fibre reinforced thermoplastics I shall consider a single family — carbon fibre reinforced polyether ether ketone. This is the only material for which extensive data are available in the open literature and detail of the same polymer as a short fibre reinforced system has already been covered in Chapter 1.[24] This system is known in two forms: film stacked sheet prepared by Hartness of the University of Dayton,[4] and by Lind of Rolls Royce;[23] and as a pre-impregnated tape, from which laminates may be constructed, as Aromatic Polymer Composite APC-1.[18] As well as the literature currently available, a number of papers considering detail of the mechanical properties of this system are in preparation.[25, 26]

## 4.4. THE MICROSTRUCTURE OF COMPOSITES BASED ON CONTINUOUS COLLIMATED FIBRES IN CRYSTALLINE MATRICES

The elements of a composite material are the reinforcement, the matrix and the interface between the two. However, the properties of a composite depend not only upon those elements, but also upon their organisation into a structure.[27]

Aromatic Polymer Composite, APC-1, is prepared as pre-impregnated tape about 0·13 mm thick containing highly collimated, high-strength carbon fibres. Microscopy (Fig. 4.1) reveals firstly, that there are no detectable voids in the material, and secondly, that the fibres are well wetted by the resin and distributed in a random array without significant bundles or extensive fibre-free areas. Because of good wetting of the fibre by the resin, lamination of the prepreg tapes to form mouldings requires only heat and modest pressures ($10^6$ N/m$^2$ is recommended[28]), and that lamination process does not lead to any serious disruption of the fibre array.

The resin phase of APC-1, polyether ether ketone, is a crystallisable polymer[29] and, as with any crystallisable polymer, the detailed microstructure of the resin phase depends on thermal history between the glass transition temperature, 143 °C, and the melting point, 343 °C. Strategies for rapid fabrication of finished parts require rapid heat transfer but, for structural engineering materials, it is essential that the morphology of the product should be uniform and reproducible. Blending these two requirements for rapid fabrication and uniform microstructure requires careful control of the dynamics of crystallisation

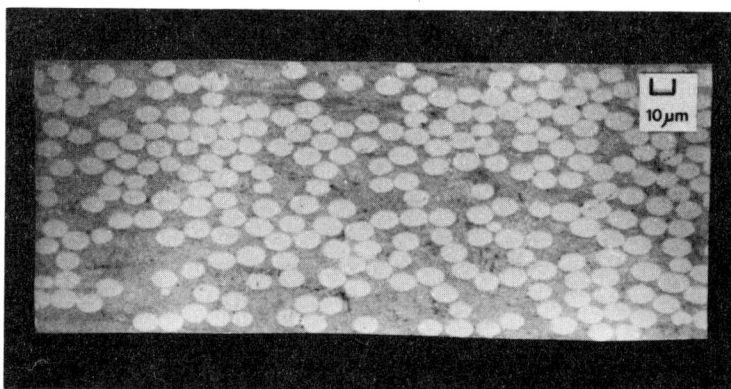

**Fig. 4.1.** A micrograph of APC-1 prepared from a pre-impregnated tape.

and with APC-1 we are able to obtain reproducible structures provided that the cooling rate for the product from 340 °C to 200 °C is in the range 40–2000 °C/min. This 50-fold range permits the process designer access to a wide variety of high-speed production technologies, while allowing the product designer to maintain the full benefit of controlled crystallinity. A complementary grade, currently under development, is especially suitable for moulding thick sections and allows cooling rates in the range 'slow' to 400 °C/min.[30]

The third factor in the microstructure of a composite material is described by Drzal[31] as the 'interphase' concerned with the detail of the interaction between the resin and the fibre at the boundary between them. In the case of thermosetting carbon fibre/epoxy systems, Drzal concludes that this interphase is likely to be more brittle than the base resin phase. McMahon[32] has indicated how control of that interphase, essentially by permitting debonding and fibre pull out, can be used as a tool to modify the toughness of the composite structure with thermosetting polymers. In Aromatic Polymer Composite APC-1, the interphase is different from that encountered in thermosetting systems. In these materials there is the potential for the resin to crystallise directly on to the fibre, thereby forming a continuous structure with it so that even when the composite is broken that bond remains intact.[27]

The first requirement of an optimum design is reproducibility of the material from which the product is to be made. The key to reproducibility is control of the microstructure.

## 4.5. THE MACROSTRUCTURE OF CONTINUOUS FIBRE REINFORCED COMPOSITES

Above the microstructure there is another level of structure in composite materials, one which is under the control of the product designer. This is the orientation of layers of fibres within a structure. Where the optimum of continuous collimated fibres is employed, as is the case with Aromatic Polymer Composite APC-1, the orientation of each layer of a laminated structure is readily defined with respect to a control axis. To identify the construction of such a laminate we indicate the ply orientation with respect to the control axis and the number of such laminates. Thus $0_4$ indicates that four layers of pre-impregnated tape are laminated together with their axis parallel to the control axis (Sketch 4.1). By contrast $90_4$ implies four layers with their axis transverse to the control axis (Sketch 4.2).

A set of brackets ( ) indicates a repeated sequence while the numerical suffix indicates the number of times that the sequence is repeated. The suffix S indicates that the lay up is then symmetrical about the central axis. $(0,90)_{2S}$ could thus be rewritten 0,90,0,90,90,0,90,0 (Sketch 4.3).

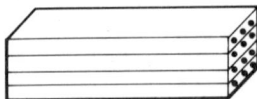

Sketch 4.1.

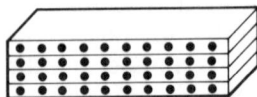

Sketch 4.2.

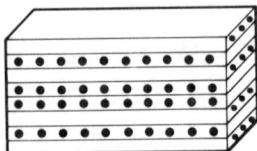

Sketch 4.3.

Symmetry about a central axis is of very great importance in the construction of laminates from highly anisotropic products such as prepreg tape, otherwise thermal stresses may cause buckling. In the case of a symmetric lay up involving an odd number of plies the central ply which is not repeated is indicated by a bar; thus $(0,90,\bar{0})_S$ indicates a five-layer stack 0,90,0,90,0.

A so-called quasi-isotropic laminate is a common form which has approximately uniform properties in all directions in the plane of the sheet. Popular forms of such laminates include $(+45, 90, -45, 0)_{NS}$ and $(0, 90, +45, -45)_{NS}$ where $N$ is any number.

Continuous collimated fibre may also be woven into a sheet and many studies of 'film stacked' materials are based on direct impregnation of a woven cloth. For convenience we identify a woven cloth as 0/90 or $\pm 45$. Thus $(0/90, \pm 45)_S$ and $(0, 90, +45, -45)_S$ are approximately equivalent quasi-isotropic lay ups in woven and simple laminates respectively. Woven broad goods may also be prepared from pre-impregnated tapes.

All composites based on continuous collimated fibre possess strong, but defineable, anisotropy. In the following description of basic mechanical properties we shall define the longitudinal properties (along the fibre axis; Sketch 4.4), the transverse properties (normal to the fibre axis; Sketch 4.5), and a measurement at 45° to the fibre axis based on a symmetrical cross-ply laminate $(-45, +45)_S$ which is a strong indication of shearing within the composite (Sketch 4.6).

Sketch 4.4.

Sketch 4.5.

Sketch 4.6.

Further, we need an indication of shearing within a laminate along the fibre axis (Sketch 4.7), transverse to the fibre axis (Sketch 4.8), and indeed between layers at different orientation (Sketch 4.9).

Sketch 4.7.

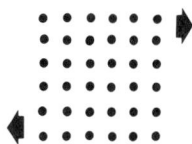

Sketch 4.8.

Sketch 4.9.

Finally we need to obtain some indication of the resistance of the laminate to a peeling force, illustrated in Sketch 4.10.

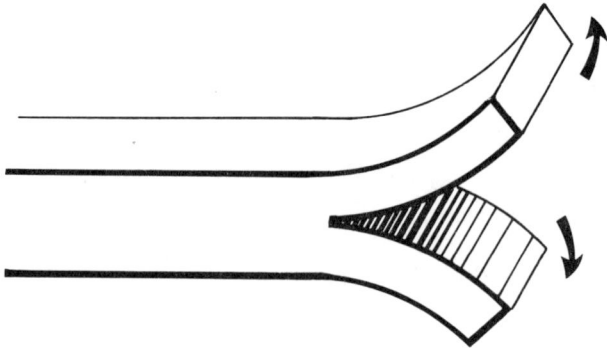

Sketch 4.10.

## 4.6. THE MECHANICAL PROPERTIES OF CARBON FIBRE REINFORCED POLYETHER ETHER KETONE (PEEK)

### 4.6.1. Stiffness

The carbon fibre used in Aromatic Polymer Composite APC-1 has a modulus of 237 GN/m$^2$ while the resin (PEEK) has a modulus of 4 GN/m$^2$.

In a simple tensile test with the loading along the fibre axis the resin and fibre share the load applied in parallel (Sketch 4.11). The law of mixtures determines the potential axial stiffness of a fibre reinforced composite along the fibre axis as

$$\text{Axial Tensile Stiffness} = \sum v_i E_i$$

where $v_i$ and $E_i$ are the volume fraction and modulus of the $i$th component. All elements are subject to the same strain.

**Sketch 4.11.**

The theoretical tensile modulus of APC-1, containing 0·52 volume fraction of fibres, along the fibre axis is 125 GN/m$^2$ and this theoretical value should be achieved if each fibre is thoroughly wetted and connected to its neighbour so that it can share its proper proportion of the load.

By contrast, if the load is applied transverse to the fibre axis we may make a first approximation to the components being loaded in series (Sketch 4.12). Here each element is subjected to the same stress. Then

$$\text{Transverse Tensile Compliance} = \sum v_i J_i$$

**Sketch 4.12.**

where $v_i$ and $J_i$ are the volume fraction and compliance of the $i$th component. From this argument we may deduce an approximate tensile stiffness for APC-1 normal to the fibre axis of 8 GN/m$^2$. This approximation ignores the very severe constraints under which the thin compliant layers of resin are placed, and a slightly higher value should actually be recorded.

The axial and transverse values of stiffness mark the upper and lower bounds. Theoretical computation of off-axis stiffness involves complex analysis, but highly developed theories do exist.[33] On the basis of these theories we may estimate the stiffness of an individual ply of APC-1 oriented at an angle of $\theta$ to the loading axis as illustrated by Fig. 4.2, which also projects the modulus for plies laid up at $\pm \theta$.

A common practical lay up is designed to give an approximately balanced property in all directions in the plane of the laminate — an organisation usually described as quasi-isotropic. While the precise calculations of laminate theory[33] are appropriate for such structures a useful first approximation is that the in-plane stiffness of a quasi-isotropic plaque is one-third of the stiffness of a uniaxial laminate. In

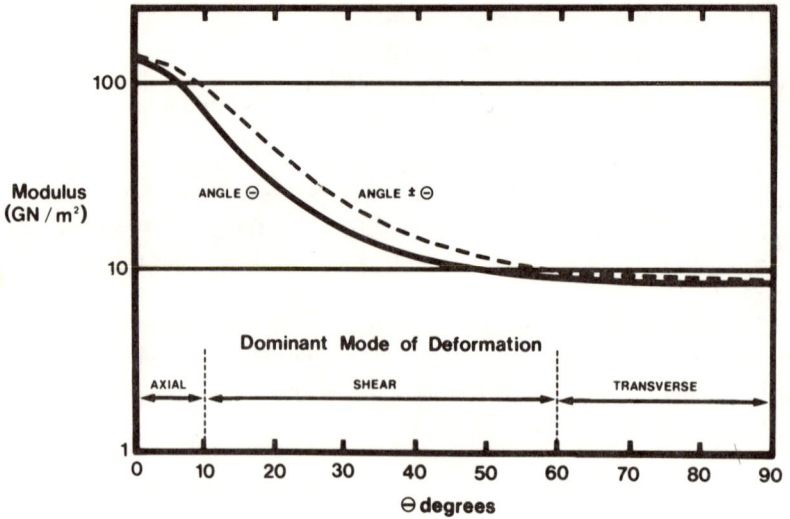

**Fig. 4.2.** The stiffness of an individual ply of APC-1 orientated at an angle $\theta$ to the loading axis.

the case of APC-1 we would thus anticipate the stiffness of a quasi-isotropic plaque to be about 40 GN/m².

At this point we must introduce experimental evidence so that we can judge if the system is indeed conforming to the expectation of a well-impregnated fibre reinforced plastic. Further we need to introduce experimental determinations of Poisson's ratio as an aid to identify the constraints on layers in a full-scale laminate structure. Analysis of the data displayed in Table 4.1[34] allows us to deduce the in-plane shear modulus along the fibre axis, and torsion[35] testing permits us to deduce the shear modulus transverse to the fibre axis.

TABLE 4.1
Tensile Testing at 23 °C

| Sample lay up | Modulus[a] (GN/m²) | | Poisson's ratio in the plane[a] | |
|---|---|---|---|---|
| $0_{16}$ | 122 | (8) | 0·31 | (0·02) |
| $(+45, -45)_{4S}$ | 14·2 | (0·3) | 0·75 | (0·01) |
| $90_{16}$ | 9·2 | (0·9) | 0·038 | (0·005) |

[a]Standard deviation is quoted in parentheses.

TABLE 4.2
Shear Properties at 23 °C

| *Direction* | *Modulus* *(GN/m²)* |
|---|---|
| Axial | 4·5 |
| Transverse | 4·0 |

By comparing the axial and transverse moduli (Table 4.1) with the theoretical values estimated in Fig. 4.2 we deduce a very high efficiency of wetting, reinforcing the conclusion of our study of the microstructure (Fig. 4.1). This conclusion is of particular importance since it permits us to make full use of the carefully constructed theoretical laminate analyses available for such systems.

For engineering design purposes a simple study of stiffness at short time scales at ambient temperature is only partially adequate. We require data on the mechanical response at long times and at the range of temperatures likely to be experienced in practice.

Based on studies of Dynamic Mechanical Analysis[36] we can deduce that the stiffness of APC-1 does not significantly vary in the temperature range −160 °C to +120 °C. However, the resin undergoes a glass transition process at 143 °C and above that temperature the stiffness falls significantly in a lay up such as ± 45° where the properties are 'resin dominated'. Despite this significant drop the stiffness of a uniaxial 'fibre dominated' orientation is little affected by the glass transition process and, because the resin is crystalline, a very respectable balance of stiffness properties is retained until close to the melting point (Fig. 4.3).

Thermoplastic resins tend to have a 'bad name' for creep resistance based largely on experience with resins such as polyethylene which were not originally designed for structural materials. However, Hartness[4] has demonstrated that carbon fibre reinforced polyether ether ketone has superior creep resistance to 'state of the art' carbon fibre reinforced epoxy systems. Thus, in respect of axial loading, APC-1

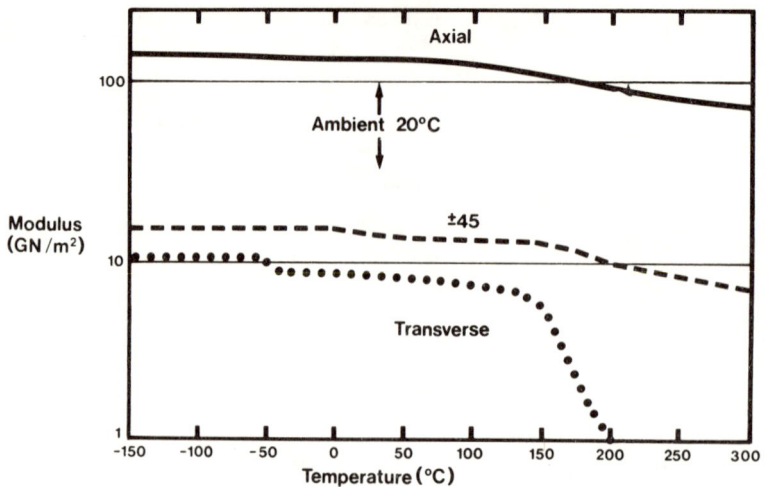

**Fig. 4.3.** The variation of modulus with temperature for APC-1.

demonstrates virtually no creep[37] and even when loaded transverse to the fibre axis — the worst possible configuration — the creep response is low[38] (Fig. 4.4).

In respect of stiffness we may conclude that a fibre reinforced thermoplastic can:

(i)   by being fully wetted, conform to the expectations of fundamental theoretical analysis;

(ii)  because of the inherent properties of the resin, offer virtually invariant properties over the temperature range −160°C to +120°C and at extended time scales;

(iii) by virtue of crystalline structure, offer a potentially useful range of properties well above the glass transition temperature.

### 4.6.2. Strength

The fibre used in APC-1 has a tensile strength of 3430 MN/m² whilst the tensile yield stress of the resin is 100 MN/m². The concept of strength is rather less amenable to fundamental analysis than the theoretical arguments about stiffness except in the case of simple tensile strength of a uniaxially aligned fibre, where the law of mixtures should apply. According to the law of mixtures the simple tensile strength of APC-1 based on 52% by volume of carbon fibre is 1790 MN/m². The tensile

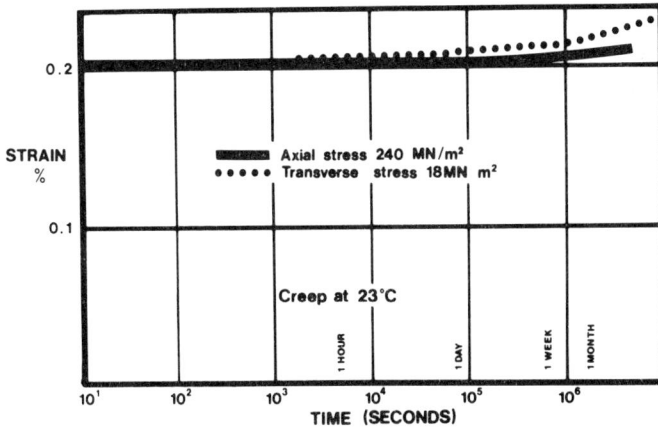

**Fig. 4.4.** Creep curve for APC-1 at 23 °C.

strength of a 2 mm thick laminate of APC-1 conforms to that prediction (Table 4.3).

TABLE 4.3
Tensile Strength of APC-1 at 23 °C

| Sample lay up | Strength[a] $(MN/m^2)$ | Strain at failure (%) |
|---|---|---|
| $0_{16}$ | 1 830 (110) | 1·5 |

[a]Standard deviation is quoted in parentheses.

The fact that the full inherent strength of the fibre can be realised in a section some 300 fibres thick is an important factor in confirming good wetting of the fibres, good collimation of the fibres within a layer and satisfactory bonding between the constituent laminates.

A measurement of tensile strength transverse to the fibre orientation should be capable of giving a value approximating to the tensile strength of the resin, except that the fibres represent a severe constraint on the resin which may prevent it from developing its full strength. In practice we have not been able to make a direct measurement of transverse tensile strength because of the tendency to break at the additional constraint provided by the clamp: those measurements indicate that the transverse tensile strength of APC-1 at 23 °C is greater than 65 MN/m$^2$.

On samples loaded at an angle to the fibre axis the inherent ductility of the resin permits a shear yielding phenomenon in APC-1 which appears to be unique in continuous fibre composite materials (Fig. 4.5; Table 4.4).[38] Note that the strain at failure of 21% represents a readily visualised distortion of the sample, as depicted in Sketch 4.13, emphasising the potential for such a structure to absorb a considerable amount of energy before failure.

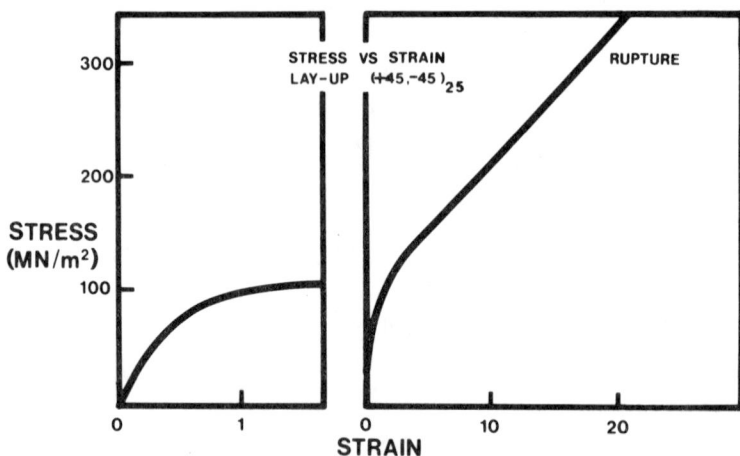

Fig. 4.5. A shear yielding in APC-1.

TABLE 4.4
Tensile Strength of APC-1 at 23 °C

| Lay up | Stress at failure[a] (MN/m$^2$) | Strain at failure (%) |
|---|---|---|
| (+45, −45)$_{4S}$ | 340 (19) | 21 |

[a]Standard deviation is quoted in parentheses.

It is easy to visualise how fibres can offer strength in tension but, in compression, there is a natural tendency for the fibres to buckle unless they are supported by close mutual packing and supported by a stiff resin. In APC-1 both the fibre content (52% by volume) and the resin

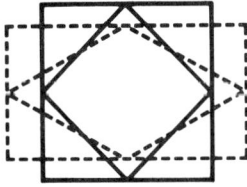

Sketch 4.13.

modulus (4 GN/m$^2$) are low to achieve the optimum strength of the fibre in compression. Indeed the very ability of the resin to yield, while enhancing the toughness of the composite, may slightly compromise the compression strength. Thus, in general, compression strength is lower than tensile strength, as demonstrated in Table 4.5.

TABLE 4.5
Typical Strength of APC-1 at 23 °C

| Lay up | Tensile strength (MN/m$^2$) | Compression strength (MN/m$^2$) |
|---|---|---|
| Uniaxial | 1 830 | 950 |
| Quasi-isotropic | 700 | 350 |

Because of its great convenience as a laboratory test, and also for its considerable practical significance, flexural testing is a valued tool for the assessment of materials. Flexure, as typified by the three-point bending test, is simple to carry out but complex to analyse. The convex surface of a flexure specimen is under tension, the concave surface is in compression, while the central regions are subject to shearing stresses (Sketch 4.14). The relative magnitude of these stresses depends on the geometry of the test. If the specimen is long and thin then the shearing stresses are of little significance, but because of the very great difference between tensile and shear modulus a span-to-depth ratio of at least 40:1 is recommended when testing along the fibre axis. If the specimen is short and fat then shearing stresses will predominate: a span-to-depth ratio of 5:1 is commonly used to give a shear dominated deformation. There is one further complication in flexural testing of composite materials; because the stiffness is high, the applied loads are high and the load, being applied at the point of maximum stress, may damage the top surface leading to premature compression failure. This problem can be avoided by the use of a loading bar of sufficiently large radius of

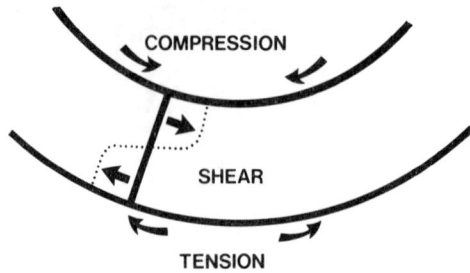

Sketch 4.14.                          TENSION

curvature. Before drawing any scientific conclusion from flexural testing the geometry of the test must be carefully analysed. Typical values of stiffness and strength for APC-1 are displayed in Table 4.6.

TABLE 4.6
Typical Values of Stiffness and Strength Derived from Flexural Testing at 23 °C on Specimens 2 mm Thick Using a 6·35 mm Loading Bar

|  | Axial | | Transverse | | Shear |
|---|---|---|---|---|---|
| Specimen | Modulus $(GN/m^2)$ | Strength $(MN/m^2)$ | Modulus $(GN/m^2)$ | Strength $(MN/m^2)$ | Strength $(MN/m^2)$ |
| Uniaxial | 120 | 1 670 | 10 | 110 | 106 |
| Quasi-isotropic | 47 | 700 | 47 | 700 | 80 |

### 4.6.3. Toughness and the Concept of Damage Tolerance

Of all the properties of materials, toughness is perhaps the least readily defined and yet the most critical in terms of confidence in a structural component. The difficulty of fundamental definition is productive of theories, of elegant test methods and of intellectual argument but it means that in the end we must always have recourse to a test which simulates service conditions. In practice this usually means bashing a sheet on the corner of a filing cabinet and then seeing what damage has been done.

The need for a laboratory test has produced a range of analytical versions of the filing cabinet test of which the Instrumented Falling Weight Impact Test[39] is one informative option. A sheet specimen is placed on a circular support and impacted in the centre with a dart of specified geometry and known mass and velocity. As the dart penetrates the sheet the force is recorded. This test gives three values which can be used to characterise the toughness of the sheet (Fig. 4.6). From high-speed photography[40] it appears that the *peak force* corresponds to the first breaking of fibres on the back surface. We may then integrate the area under the force deflection curve to give an *initiation energy* for failure. The area under the completed curve at burst through is then defined as the final *failure energy*.

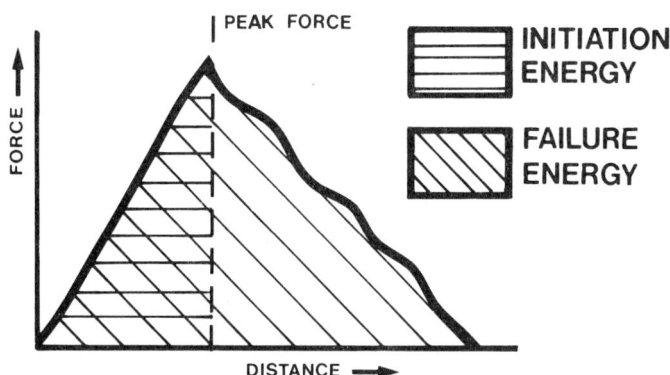

Fig. 4.6. A typical result from an Instrumented Falling Weight Impact Test.

Under standard testing conditions (diameter of support ring, and the geometry, mass and velocity of the dart) the main variable is sample thickness. Typical results are shown in Table 4.7. Empirically, we deduce that peak force, and initiation and failure energies all vary with (thickness)$^{1.4}$ — a result which is also found for a wide range of other materials tested in this way.[38]

TABLE 4.7
Instrumented Falling Weight Impact Test on Quasi-isotropic laminates of APC-1 at 23 °C

Support diameter 50 mm, dart diameter 12·7 mm, velocity 5 m/s, mass 5 kg. Standard deviation quoted in parentheses, five tests on each sample.

| Sample thickness (mm) | Peak force (N) | | Initiation energy (J) | Failure energy (J) |
|---|---|---|---|---|
| 0·66 (0·01) | 914 | (108) | 1·8 (0·2) | 4·0 (0·3) |
| 1·15 (0·02) | 1 629 | (113) | 3·7 (0·5) | 8·6 (0·6) |
| 2·17 (0·06) | 3 672 | (287) | 7·8 (1·1) | 20·9 (1·8) |
| 3·58 (0·07) | 7 242 | (334) | 16·8 (1·3) | 44·8 (3·7) |
| 5·84 (0·08) | 14 130 (1 010) | | 37·8 (3·2) | 98·7 (3·6) |

Factors such as temperature would also be expected to have a significant effect on impact resistance, but for APC-1 only slight variations are seen within the range of likely service temperature (Table 4.8).

TABLE 4.8
Impact Resistance of APC-1 of Thickness 1·35 mm

| Temperature (°C) | Peak force[a] (N) | Initiation energy[a] (J) | Failure energy[a] (J) |
|---|---|---|---|
| −50 | 1 582 (118) | 3·4 (0·3) | 9·3 (0·3) |
| +23 | 1 503 (147) | 3·7 (0·7) | 10·3 (2·0) |
| +90 | 1 709 (97) | 5·1 (0·7) | 11·9 (1·5) |

[a]Standard deviation is quoted in parentheses.

Closer inspection of samples subject to low energy impacts, for example using ultrasonic C-scan techniques and microscopic sections, indicates that some damage has been caused within the sample before the peak force is reached. That damage is in the form of cracks in transverse plies and delamination between the plies (Fig 4.7).

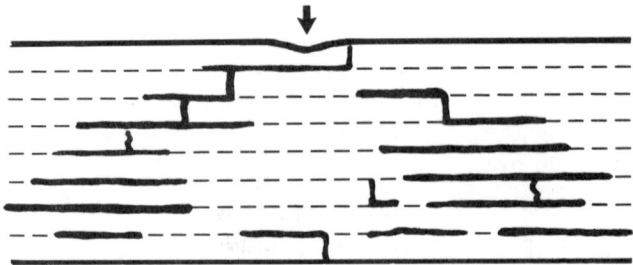

Fig. 4.7. Cone of damage in impacted sample.[41]

There are two views about delamination. If a sample suffers extensive delamination, then a large amount of new surface is created and that requires energy: high delamination and high energy absorption go hand in hand and increasing delamination and debonding of the fibre[32] is one way of enhancing energy absorption in a composite structure designed specifically as a 'one-shot' shock absorber. The alternative argument is that any partial delamination seriously compromises subsequent strength in shear or compression loading. This latter view prevails amongst designers who wish to ensure their structure against adventitious damage and has produced a range of tests based on measuring strength after a prescribed damage. Typical delamination under impact for Aromatic Polymer Composite APC-1 is represented graphically in Fig. 4.8 together with residual strength in compression measured according to one particular test.[42]

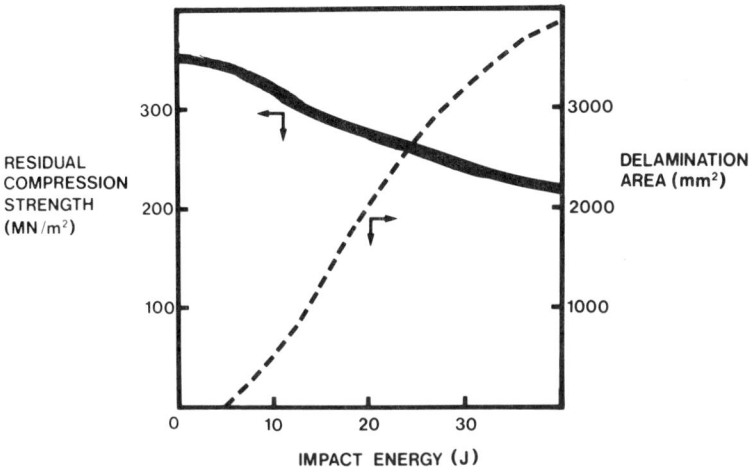

Fig. 4.8. Delamination and residual strength for APC-1 at 23 °C.[38] Sample lay up $(+45, 90, -45, 0)_{4S}$; thickness 4·5 mm.

In the Instrumented Falling Weight Impact Test the delamination area after punch through is, like the energy, a function of sheet thickness: for APC-1 the delamination area increases with (thickness)$^{0.7}$.

In respect of damage tolerance it is the resistance to delamination which is a major factor: we must therefore seek a further appreciation of the resistance to the spread of delamination under load.

The resistance of composites to the propagation of interlaminar cracks has been the subject of rigorous scientific study based on tests such as the double cantilever beam represented in Sketch 4.15,[43] where the energy required to propagate a crack between two plies can be directly measured. Hartness,[4] working with film stacked woven carbon fibre/PEEK, obtained an interlaminar strain energy release rate, $G_{1c}$, of 1300 J/m$^2$, a value which he observed was one order of magnitude greater than that found in 'state of the art' thermosetting resins. With uniaxial composites of APC-1, Carlile and Leach[44] recorded values between 1800 and 3200 J/m². This resistance to delamination appears to be associated with the ductile character of the resin which can take many forms, depending on the peel rate, from gross yielding to microductility. An example of this is shown in Fig. 4.9.

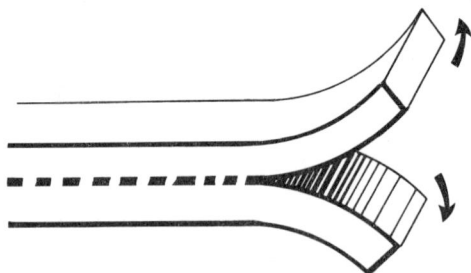

Sketch 4.15.

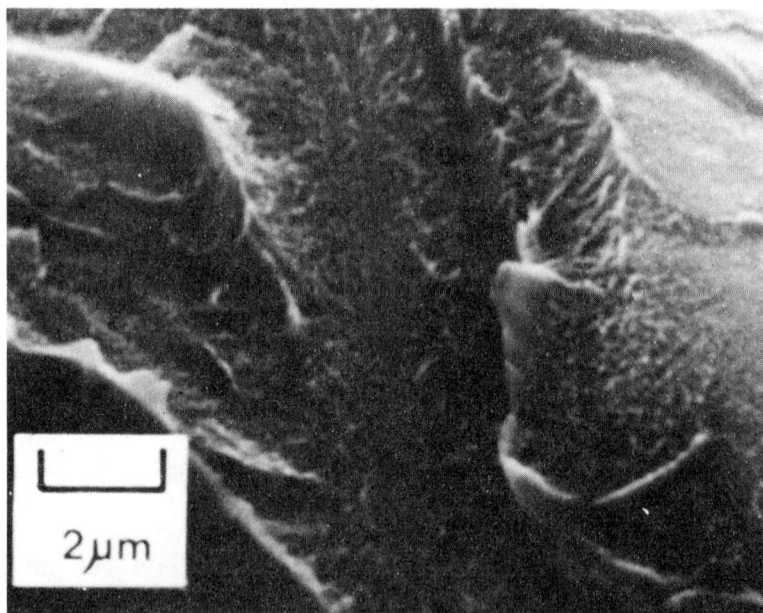

Fig. 4.9. A scanning electron micrograph showing the exposed fibre surface with adhering resin.

Carlile[44] identifies the high resistance to crack propagation as one reason for good damage tolerance as exemplified by the retention of compression strength after a component has been subjected to impact. This aspect of toughness is probably also the controlling factor in the enhanced fatigue resistance observed by Hartness[4] using film stacked materials.

That the inherent toughness of the thermoplastic resin can be carried through to the composite and be manifested in service is far from a foregone conclusion. A not-uncommon experience of the simple addition of short fibre reinforcement to an inherently tough thermoplastic matrix is to produce a stiff moulding which is likely to shatter on impact. The translation of matrix characteristics into composite performance is fraught with difficulties. The two most obvious problems are the geometry of the resin phase and the constraints imposed on the resin by the fibres.

In our consideration of the microstructure of composite materials we identified that the mean thickness of the resin phase was only about 2·5 μm and particle sections show it varying down to below 1 μm (Fig. 4.1). The teachings of fracture mechanics suggest that thin layers will be tougher than thick ones and this has led to the proposal of a criterion for toughness in composites based on comparing the plastic zone size of the resin phase with the characteristic thickness of the resin.[27] However, that criterion is beset with difficulties since those thin layers are likely to be subjected to intense thermal stresses induced during fabrication. A simple way to demonstrate such stresses is to lay up an unbalanced sheet such as 0,90 and observe that it is very difficult to make it lie flat. Not only does the moulding warp but the warp is bistable — a pleasing executive toy to relax the fingers. By using continuous collimated fibres the analyst is at least spared worries about the stress concentrations at fibre ends and at loops or crossovers: the elimination of those stress concentrations undoubtedly helps the material to realise its full toughness. The micromechanical description of the state of stress in a composite material, and how it defines toughness, is a scientific story which has yet to be written.

### 4.6.4. Ageing and Environmental Resistance

Possession of a good basic property profile is only the first hurdle which a high-performance composite must clear. The subsequent hurdles require the retention of those properties in service and, in particular, in the presence of hostile environments. The definition of hostile environment, and the test method most appropriate to that environment, depends on the service to which the composite is destined: each user must necessarily specify his own testing strategy. We can, however, define four basic categories: ageing, water, solvents and fire.

In respect of ageing resistance with a product as new as APC-1, we must rely on accelerated testing. This material has been designed for continuous service between −50°C to +80°C. The fibres themselves may be presumed to be insensitive to ageing and it is thus the resin and interface which we must test for potential problems. As with creep properties, the resin and interface characteristics can be exaggerated by measurement transverse to the fibre axis. We have used a transverse flexural strength after ageing at 120°C and 220°C (Table 4.9): these temperatures are selected as those at which free volume and crystallisation changes are to be most expected. The samples were kept in an air environment so that oxidation might also occur.

TABLE 4.9
Transverse Flexural Strength[a] (MN/m$^2$) of APC-1 at 23°C

| Ageing temperature (°C) | Transverse flexural strength | | | | | | |
|---|---|---|---|---|---|---|---|
| | 0 h | 4 h | 8 h | 48 h | 216 h | 504 h | 1 320 h |
| 120 | 147 | 132 | 141 | 136 | 140 | 154 | 144 |
| 220 | 147 | 137 | 138 | 150 | 132 | 120 | 103 |

[a]Coefficient of variation 7%.

We anticipate that the decrease in strength after 1000 h at 220°C may be associated with oxidation. It must be stressed that this is an *accelerated* test from which we deduce that under normal service conditions no significant ageing effects should be expected.

A major preoccupation of the recent literature of composites is the issue of resistance to water, especially at elevated temperature. Working with film stacked material, Hartness[4] has demonstrated that this problem is eliminated by the use of polyether ether ketone resin. This result is substantiated in the case of APC-1, as shown in Table 4.10.

TABLE 4.10
Flexural Testing on Uniaxial Laminates of APC-1

| Temperature (°C) | Flexural modulus (GN/m²) | Flexural strength (MN/m²) | Short beam shear strength (MN/m²) |
|---|---|---|---|
| 23 | 119 | 1 400 | 97 |
| 95 dry | 126 | 1 390 | 79 |
| 95 wet[a] | 131 | 1 427 | 83 |
| Standard deviation | 3 | 100 | 3 |

[a]After 7 days water at 95 °C.

Note, however, that polyether ether ketone resin of itself is not sufficient to protect the composite if the interface between the resin and the fibre is not correctly established.[18]

In respect of solvent resistance, Hartness[4] has identified polyether ether ketone as having good resistance to the reagents of interest in aerospace applications, a result confirmed by Rolls Royce.[45] The only common solvent which will attack polyether ether ketone is concentrated sulphuric acid. Paint strippers are almost by definition, one of the most agressive reagents, but have no detected effect on APC-1 (Table 4.11).[46]

TABLE 4.11
Resistance of APC-1 to Paint Stripper

| Specimen | Flexural properties at 23°C | | |
|---|---|---|---|
| | Flexural modulus (GN/m²) | Flexural strength (MN/m²) | Short beam shear strength (MN/m²) |
| Control | 119 | 1 400 | 97 |
| After one month in 'Ardrox' 2526 | 124 | 1 480 | 97 |
| Standard deviation | 3 | 100 | 3 |

Like most other materials, polyether ether ketone is consumed in a flame; however, it has a high char yield, is self-extinguishing and has a low toxic gas evolution.[28] In comparison with sheet aluminium, APC-1 composite sheet is slower to heat up (because of lower thermal conductivity) and much slower to burn through (Table 4.12).[47]

TABLE 4.12
Fire Resistance[a] of Sheet 1·5 mm Thick

| Specimen | Time to red hot (min) | Time to burn through (min) |
|----------|:---------------------:|:--------------------------:|
| Aluminium | 2 | 6 |
| APC-1 | 6 | >15 |

[a]5 cm diameter flame at 1 000°C.

Environmental resistance is mainly determined by the properties of the resin and by the ability of resin to protect the interface between the matrix and the reinforcement. The very wide range of thermoplastics available enables the user to select a resin system for particular environments; the especial properties of polyether ether ketone make it a candidate for demanding aerospace applications.

## 4.7. FABRICATION STRATEGIES, REPAIR AND RECLAIM

All of the properties which we have so far identified are obtained on simple laminate structures produced by compression moulding under pressure of $10^6$ N/m$^2$ (10 atmospheres) for about 10 min. However, because the 'prepreg' is fully wetted it actually only requires nominal pressures for a few seconds to establish a consolidated laminate: polyether ether ketone is an excellent hot melt adhesive. To be of service the prepreg must be capable of being transformed into a shaped product. Brewster and Cattanach[48] identify two major classes of fabrication technology especially suited to this class of material: automated lay up and adaptations of metal working technology.

In the field of automated lay up, filament winding, and its big brother tape laying, can be carried out with thermoplastic 'prepreg' tape. In such a process the tape need only be melted for a brief time and then solidified. This ability to achieve a form-stable product within seconds of laying down means that the process designer need not be constrained to a geodesic winding pattern, neither need he feel constrained by the thought of the size of the autoclave available to 'cure' the moulding: with thermoplastics there is no curing stage.

Whatever we, as plastics technologists, may say about the economic advantages of plastics processing, the pre-eminent technology for moulding large area parts remains 'metal bashing'. Brewster and Cattanach[18] found that certain metal working technologies like roll forming and hydroforming may be readily adapted for use with continuous fibre reinforced thermoplastics. This adaptation involves heating up a preformed laminate, for example by infrared heating, and transferring it to a cold shaping tool. In the case of intermittent processes like hydroforming, a total cycle time — melting, transfer, shaping and cooling — can be less than 2 min, while for continuous processes like roll forming speeds of 10 m/min are attainable. Although the continuous fibre reinforcement constrains deep drawing, the addition of fabric technology permits a wide variety of shapes, including two-dimensional curvature, to be produced.

Composite materials are reinforced glue. If the glue is a thermoplastic, then its capacity to stick is not impaired by its having once been used. Advanced composite structures must be subjected to rigorous scrutiny to ensure that they have been correctly fabricated; in particular, delamination may be assessed by ultrasonic testing. If the inspection processes reveal any inadequacy of consolidation in the product, it is possible to remelt the composite and reform it; such an operation may, if necessary, be carried out a number of times. This ability to reconsolidate is potentially of significant economic importance, especially if rigorous quality control is demanded on highly complex mouldings. The thermoplastic character of the resin also permits a secondary, different, forming operation to be applied — an area of significant importance if a partially or fully enclosed shape is to be formed. The advantage of post-forming may be especially valuable where, for space saving reasons, it is convenient to transport a material as simple sheet stock before it is formed into sections for its final application. This approach is demonstrated by the concept of the Grumman 'beam builder'[49] for space station construction. Further, the thermoplastic characteristic can be used to form a bond between two composite components, between a component and other substructures, or to ensure a good bond when a continuous fibre reinforced element is used to give specific reinforcement to a standard injection moulded product.[50] These are new strategies for composite material fabrication which depend on the thermoplastic character of the glue phase; in particular, we note that polyether ether ketone is a good hot melt adhesive.

The potential for post-forming offers opportunities to develop a new approach to the repair of composite materials. Although we have stressed the ability of these structures to tolerate damage we must accept the inability to design out the enthusiastic fork lift truck driver, and further must acknowledge that a major market for high-performance composite systems is the military field in which battle damage must be accommodated. In both these areas we require the ability to repair damage. Two classes of damage can be identified: delamination and fibre breakage. A sheet of Aromatic Polymer Composite APC-1 subjected to repeated low-energy impacts can suffer severe delamination which adversely affects its subsequent stiffness and strength. If the impact energy is increased to a level where the fibres are broken, then, of course, residual strength and stiffness are further decreased — ultimately to zero. Any delaminated sheet can be reconsolidated by a simple remoulding process to regain the full property profile, and in the case of the worst possible fibre breakage some 50% of the original properties can be recovered:[28] Fig. 4.10 illustrates typical results. These

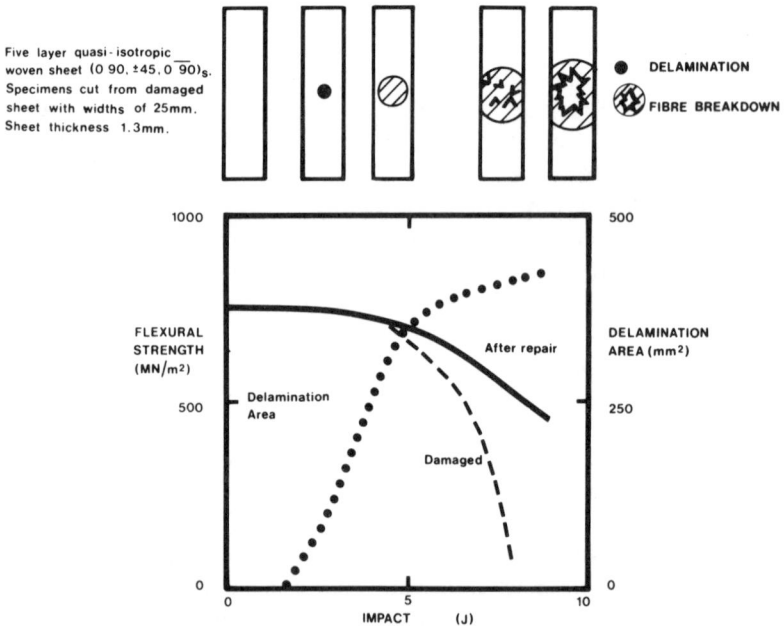

**Fig. 4.10.** The effect of reconsolidation on the mechanical properties of damaged APC-1.

repair studies were carried out under laboratory conditions; a useful field repair can, in principle, be carried out with a heavy-duty soldering iron.

If, in the design of a product, we are to make the greatest possible use of continuous fibre reinforcement, it is almost inevitable that there will be offcuts and other scrap. The economics of fabrication demand a minimisation of scrap and also an attempt to reclaim if possible. Further, no matter how high-technology a product is, it has a finite lifetime, after which its components may be recovered for other uses. Continuous fibre reinforced thermoplastics are more than cousins of short fibre reinforced thermoplastics. The relationship is such that, if they are no longer of service, continuous fibre reinforced thermoplastics may be chopped up, diluted with additional resin, and translated into injection moulding compounds — but not standard injection moulding compounds. In standard injection moulding materials, no matter what length of fibre is fed to the compounding machine, the fibres which end up in a moulding rarely exceed 0·5 mm in length. A properly wetted continuous fibre reinforced thermoplastic may be chopped to length and fed through an injection moulding process with little attrition because the individual fibres are protected from abrading each other by the viscous resin between them. Thus it is possible to injection mould chopped continuous fibre reinforced thermoplastic to obtain a product with enhanced properties, especially in respect of impact resistance and notch sensitivity.[51] Scrap from APC-1, ground up and diluted with additional 'Victrex' PEEK resin, may be reclaimed as a premium grade injection moulding composition (Table 4.13).

TABLE 4.13
Mechanical Properties of Reclaimed APC-1 Compounds

| Specimen | | Carbon fibre (wt %) | Flexural properties of injection moulded ASTM bar | |
| --- | --- | --- | --- | --- |
| | | | Modulus ($GN/m^2$) | Bar strength ($MN/m^2$) |
| Standard compound | 'Victrex' PEEK 40 30 CA | 30 | 21 | 3·0 |
| Reclaimed scrap | APC-1 + PEEK 45G | 25 | 22 | 3·2 |
| | | 50 | 41 | 5·9 |

## 4.8. THE FUTURE OF CONTINUOUS FIBRE REINFORCED THERMOPLASTICS

Commercial continuous fibre reinforced thermoplastics have undergone a prolonged gestation period during which speculation about their potential properties has been amply supported by experience of 'film stacking' compression moulding technology. However, 'film stacking', which has been commercially operated up to a scale of 1 m² left one question unanswered — 'How are we going to make wings?'.

The commercialisation of continuous fibre reinforced thermoplastics in the form of pre-impregnated tape[20, 28] as a building block from which any other form can be constructed represents a major advance in versatility which opens new opportunities to design and fabrication. Those who design and build structures now have the tools to revolutionise the field of fibre reinforced plastics mouldings.

The new commercial products based on the impregnation of continuous collimated fibre by environmentally resistant resins are directed at the most demanding aerospace applications. They are at one extreme of the spectrum of fibre reinforced plastics, which has as its opposite pole low loadings of glass in commodity resins. Between these extremes, there are already available stampable sheet products based on moderate loadings of non-woven or intermediate length fibre. There appears to be ample opportunity to develop new combinations within this field.

We are only at the beginning of the story of continuous fibre reinforced thermoplastics.

## ACKNOWLEDGEMENTS

I am indebted to my colleagues of the Aromatic Polymer Composite development team whose work this chapter reflects. In particular I would like to thank David Leach who carried out most of the mechanical property evaluations.

## REFERENCES

1. Orton, M. L., 'A new resin system for increased productivity in fibre composite processing', *International Conference on Fibre Reinforced Composites,* Liverpool, 3 April 1984.

2. Phillips, L. N., 'The properties of carbon fibre reinforced thermoplastics moulded by the fibre stacking method', RAE Technical Report 76140; *Fabrication of Reinforced Thermoplastics by Means of the Film Stacking Technique,* HMSO, London, 1980; Phillips, L. N. and Murphy, D. J. 'Thermoplastic materials', British Patent 1 570 000, 1980.

3. Hoggatt, J. T., 'Thermoplastic resin composites', *20th National SAMPE Conference,* 1975.

4. Hartness, J. T., 'Polyether etherketone matrix composites', *14th National SAMPE Technical Conference,* 1982, 14, 26; Hartness, J. T. and Kim, R. Y., 'A comparative study of fatigue behaviour of polyether etherketone and epoxy reinforced with graphite cloth', *28th National SAMPE Symposium,* 1983, 535.

5. Hogan, P. A., 'The production and uses of film stacked composites for the aerospace industry', *SAMPE Conference,* 1980.

6. Lind, D. J. and Coffey, V. J., 'A method of manufacturing composite material', British Patent 1 485 586, 1977.

7. Smith, H. R., Continuous Production of Fiber Reinforced Thermoplastics Materials, World Intellectual Property Organization WO 83/0285, 1983.

8. Turton, N. and McAinsh, J., 'Thermoplastic compositions', US Patent 3 785 916, 1974; McAinsh, J., 'The reinforcement of polysulphones and other thermoplastics with continuous carbon fibre', *BPF 8th International Reinforced Plastics Conference,* 1972.

9. Johnston, N. J., 'Interlaminar fracture toughness of composites', *28th National SAMPE Symposium,* 1983, 502.

10. British Patent 1 167 849, 1968; British Patent 1 302 048, 1970; US Patent 4 037 011, 1977.

11. McMahon, P. E. and Maximovitch, M., 'Development and evaluation of thermoplastic carbon fibre prepregs and composites', *Advances in Composite Materials,* 1980, 2, 1663 Paris.

12. Polymer Composites Incorporated, product demonstrated at exhibition attached to *SAMPE Conference,* Anahiem, 1983.

13. Price, R. V., 'Production of impregnated roving', US Patent 3 742 106, 1973.

14. Ganga, R., 'La matrice ORGASOL dans les composites fibres', paper presented at *SITEF Conference,* Toulouse, 1983.

15. Chabrier, G., Moine, G. and Maurion, R., 'Method for the manufacture of sections made of fiber containing thermoplastic resin', World Intellectual Property Organisation WO 83/01755, 1983.

16. 'Azdel', Symalit Co. Ltd, Linzburg, Switzerland.

17. 'STX', Allied Chemical Corporation, Morristown, New Jersey, USA.

18. Belbin, G. R., Brewster, I., Cogswell, F. N., Hezzell, D. J. and Swerdlow, M. S., 'Carbon fibre reinforced polyether etherketone', *SAMPE Conference,* Stresa, 1982; *Aromatic Polymer Composite APC-1* ICI PLC, Welwyn Garden City, 1982.

19. *'Fiberod' Data Sheet,* Polymer Composites Inc., Winona, USA, 1983.

20. 'Ryton' PPS Carbon Fiber Composites, Phillips Petroleum Company, Bartlesville, USA, 1983.

21. Chabrier, G. M., 'Thermoplastic matrix unidirectional composites and their applications', *SAMPE Meeting,* Toulouse, 1983.

22. Cordell, T. M., *Thermoplastic Composite Technology Development*, AFWAL/ MLBC, Dayton, USA, 1983.
23. Lind, D., Rolls Royce Ltd, Derby, private communication.
24. Clegg, D. W. and Collyer, A. A., in *Mechanical Properties of Reinforced Thermoplastics*, D. W. Clegg and A. A. Collyer (Eds), Elsevier Applied Science Publishers, London, 1986, Ch. 1.
25. Hartness, J. T., University of Dayton, paper presented at *SAMPE Conference*, Reno, April 1984.
26. Bishop, S., Curtis, P. and Dorey, G., RAE Technical Report 84061, RAE Farnborough, 1984.
27. Cogswell, F. N., 'Microstructure and properties of thermoplastic Aromatic Polymer Composites', *28th National SAMPE Symposium*, 1983, 528.
28. *Aromatic Polymer Composite APC-1*, provisional data sheet, ICI PLC, Welwyn Garden City, 1983.
29. Blundell, D. J. and Osborn, B. N., 'The morphology of poly(aryl-ether-ether-ketone)', *Polymer*, 1983, **24**, 953.
30. Aromatic Polymer Composite (development grade), ICI PLC, Welwyn Garden City.
31. Drzal, L. T., 'Composite interphase characterisation', *28th National SAMPE Symposium*, 1983, 1057.
32. McMahon, P. E. and Taggart, D. G., 'Improved carbon fibre composite performance based on new fiber and matrix developments', *SAMPE Meeting*, Stresa, 1982.
33. Chamis, C. C., 'Prediction of fiber composite mechanical properties made simple', *35th Annual Tech. Conf.*, Society of Plastics Industry, Section 12-A, 1980.
34. Leach, D. C. and Whale, M., private communication.
35. Bonnin, M. J., Dunn, C. M. R. and Turner, S., *Plastics and Polymers*, 1969, **37**, 517.
36. Osborn, B. N., private communication.
37. Moore, D. R., private communication.
38. Leach, D. C., 'The influence of a thermoplastic on the properties of structural composites', Conference organised by the Institute of Metallurgists at the Royal Society, 1982.
39. Hooley, C. J. and Turner, S., 'Mechanical testing of plastics', *Automotive Engineer*, June/July 1979, 48.
40. Jones, D. P., private communication.
41. Based on microscopy studies by Curson, A. D.
42. Byers, B. A., *Behaviour of Damaged Graphite Epoxy Laminates under Compression Loading*, NASA Contract Report 159293, 1980.
43. Whitney, J. M., Browning, C. E. and Hoogsteder, W., 'A double cantilever beam test for characterising mode one delamination of composite materials', *J. Reinforced Plastics and Composites*, 1982, **1**(4), 297.
44. Carlile, D. R. and Leach, D. C., 'Damage and notch sensitivity of graphite/ PEEK composite', *SAMPE Conference*, Cincinatti, 1983.
45. Hall, M., Rolls Royce Ltd, Derby, private communication.
46. Cogswell, F. N. and Hopprich, M., 'Environmental resistance of carbon fibre reinforced polyether etherketone', *Composites*, 1983, **14**(3), 251.

47. Briggs, P. J., *Fire Properties of Aromatic Polymer Composite (APC-1)* ICI PLC, Corporate Fire Laboratory, Blackley, private communication.
48. Brewster, I. and Cattanach, J. B., 'Fabrication with thermoplastic Aromatic Polymer Composite', *SAMPE Meeting*, London, 1983.
49. Poveromo, L. M., Muench, W. K., Marx, W. and Lubin, G., 'Composite beam builder', *SAMPE J.*, Jan. 1981, 7.
50. Cattanach, J. B., private communication.
51. Addleman, R. L., Brewster, I. and Cogswell, F. N., 'Continuous fibre reinforced thermoplastics', *BPF 13th Reinforced Plastics Congress*, 1982.

## EPILOGUE

As is noted in the section on 'The present state of the art', this chapter was written at a time when the field was rapidly evolving. All the principles of the chapter remain valid in 1986 but much of the detail has changed significantly. All the data in this chapter refer to a material system entitled APC-1. APC-1 was a first generation material and, following discussions with composite users, was replaced by an optimised system, APC-2. Both APC-1 and APC-2 are based on polyether ether ketone resin and high-strength carbon fibre and both demonstrate a similar property profile. The major difference between these two materials is that where APC-1 contained only 52% by volume of reinforcing fibre APC-2 contains 62% by volume: as a result the fibre dominated properties of stiffness and strength are greatly enhanced.

Besides the established production grade material APC-2 based on continuous collimated fibres, development grades of composite based on woven fabric reinforcement are also becoming available. While high-strength carbon fibre and polyether ether ketone remains the preferred ingredients for these composite systems, development grade materials based on higher modulus fibres and higher temperature resins are also in view. The establishment of a long-term optimised system, APC-2, has not halted the evolutionary process for other members of this family.

The principle of the impregnation of continuous fibre by thermoplastics has also been exploited in the development of improved materials for injection moulding. The 'Verton' range of injection moulding compounds, which allows the moulding of components where the fibre retains some ten times the length normally observed in such products, provides the plastics industry with a new product form combining easy moulding with stiffness, toughness and dimensional tolerance in the finished component.

Such a rapidly developing field has a rapidly expanding literature. Much of this literature is published by the Society of Advanced Materials and Process Engineering in their Journal and Symposia. One other review, 'Thermoplastic carbon fibre composites' by Paul E. McMahon, has recently been published in *Developments in Reinforced Plastics — Volume 4*: this review provides a complementary discussion of the background to these material systems and to many of the issues, such as their manufacture, which it has not been possible to address in detail in this chapter. A further complementary review entitled 'Processing with aromatic polymer composites', prepared by J. B. Cattanach and F. N. Cogswell, is due to appear shortly in *Developments in Reinforced Plastics — Volume 5*. The implications of this new class of thermoplastic composite materials on the plastics business scene is addressed by Rob Wehrenberg in an article entitled 'New composites expand action for processors' published in *Plastics World* (December 1985, pages 39–43). We are still only at the beginning of the story of continuous fibre reinforced thermoplastics but the scale of the opportunities which lie ahead is becoming clearer.

*Chapter 5*

# The Rheology and Processing of Reinforced Thermoplastics

JAMES LINDSAY WHITE

*Polymer Engineering Center, University of Akron, Ohio, USA*

## 5.1. INTRODUCTION

The fabrication of parts from particle and chopped fiber filled plastics and elastomers is carried out by pumping, extrusion and molding of bulk polymer systems in a fluid state. A proper understanding of these processing operations requires a knowledge of the rheological properties of these systems. It is the purpose of this article to describe these rheological properties, the processing of compounds and their inter-relationships.

I begin by indicating the different classes of flows which occur in polymer processing operations. I then turn to the measurement of rheological properties of importance in such flows. The influence of particle structure and loading on the rheological behavior of compounds is then summarized. I conclude by describing the processing charac-teristics of particle–polymer melt compounds and relating these to their rheological behavior.

## 5.2. PROCESSING AND CLASSIFICATION OF FLOWS

The classes of flows which occur in processing operations may generally be divided into 'internal flows' and 'external flows'. Internal flows occur between the screw and barrel of an extruder, in dies and in moving members (the screw and barrel), or due to pressure gradients. The polymer melts adhere, or nearly so, to the steel surfaces. This makes the primary velocity gradients normal to the direction of flow, i.e. they are shearing motions. Thus if '1' is the direction of flow and '2' and '3' are orthogonal to this direction, the velocity field may be expressed as $v_1(x_2, x_3)$ in these flows.

External flows generally occur in post die forming operations such as melt spinning of fibers, tubular film extrusion and blow molding. The melt emerging from the dies is stretched by externally applied tensions and inflation pressures. The flows are elongational in character. They involve motions of form $v_1(x_1)$, $v_2(x_2)$ and $v_3(x_3)$.

Generally velocity gradients in internal flows show little variation in the direction of flow. The flows are also roughly isothermal. On the other hand, velocity gradients in external flows show major variations as they are controlled by external forces and not by constant pressure gradients or relative motions of moving machine members. As external flows occur in air, the melt rapidly cools and the process is highly non-isothermal.

In these rheological investigations, both shearing flows and elonga-tional flows will be considered. It is the former which are most important for quantitative interpretation and analysis of flow behavior. A complete rheological model with the potential of handling all types of flows requires the proper interpretation of elongational flow data as well.

## 5.3. EXPERIMENTAL METHODS OF DETERMINATION OF RHEOLOGICAL PROPERTIES

### 5.3.1. Shear Flows

Shear flows involve velocity fields of form:

$$v = v_1(x_2)e_1 + 0e_2 + 0e_3 \tag{5.1}$$

i.e. the velocity is constant in the direction of flow, but varies in the perpendicular direction. The stress fields in such flows in general include both shear stresses $\sigma_{12}$ and unequal normal stresses $\sigma_{11}, \sigma_{22}$, and $\sigma_{33}$. These are usually represented as:

$$\sigma_{12} = \eta(\dot{\gamma})\dot{\gamma} \tag{5.2a}$$

$$N_1 = \sigma_{11} - \sigma_{22} = \psi_1 \dot{\gamma}^2 \tag{5.2b}$$

$$N_2 = \sigma_{22} - \sigma_{33} = \psi_2 \dot{\gamma}^2 \tag{5.2c}$$

$$p_1 = -\tfrac{1}{3}(\sigma_{11} + \sigma_{22} + \sigma_{33}) \tag{5.2d}$$

Here $\eta(\dot{\gamma})$ is the shear viscosity, $N_1$ and $N_2$ are the principal and second normal stress differences and $\psi_1$ and $\psi_2$ are the principal and second normal stress difference functions.

There is a long history of shear flow measurements. During the nineteenth century various rotational instruments, such as coaxial cylinders,[1] torsional flow between disks[2] and capillary flow,[3] were introduced and applied to measure the viscosity of Newtonian liquids and gases. In the present century the same classes of instruments were applied to polymer solutions and later to elastomers and polymer melts. The first careful measurements of shear viscosity for bulk polymers were carried out in the 1930s with the efforts of Mooney on both coaxial cylinder[4] and shearing disk torsional flow rheometers[5] and Dillon[6] on capillary rheometers. It was discovered in this period that the shear viscosity of these systems is not constant, but is a function of shear rate. Weissenberg[7] and Mooney[4, 8] pioneered the development of analyses to determine accurately viscosity–shear rate behavior of non-Newtonian fluids from these instruments.

The first clear-cut observations of normal stresses were made by

Weissenberg[9, 10] in the late 1940s. Weissenberg and his coworkers developed an instrument based on cone–plate geometry to measure normal stresses. The first normal stresses on polymer melts were reported by Pollett and his coworkers[11] and later by King.[12]

The simplest of rheometers is the parallel plate or sandwich viscometer shown in Fig. 5.1(a). This type of instrument has been applied by several investigators[13-16] to rheological characterization of filled polymer systems, generally elastomers. The shear rates in this instrument are simply given by

$$\dot{\gamma} = \frac{V}{H} \qquad (5.3)$$

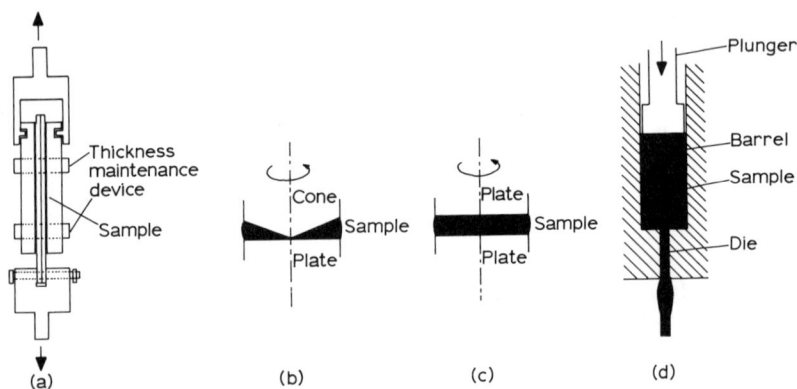

**Fig. 5.1.** Shear flow rheometers: (a) sandwich; (b) cone-plate; (c) parallel plate; (d) capillary.

where $V$ is the linear velocity and $H$, the distance of separation between the plates. The stresses are given by

$$\sigma_{12} = \frac{F(t)}{2A(t)} \qquad (5.4)$$

where $F(t)$ is a tensile force separating the plates and $A(t)$ the *contact* area.

Another instrument which has received considerable application to filled thermoplastics is the cone–plate rheometer Fig. 5.1(b).[17-29] The shear rate in this instrument is to a good approximation given by[30]

$$\dot{\gamma} = \frac{\Omega}{\alpha} \tag{5.5}$$

where $\Omega$ is the angular velocity and $\alpha$ the angle between cone and plate. The shear rate and stress components are constant through the gap. The shear stress is related to the torque $M$ through eqn (5.6).[30]

$$\sigma_{12} = \frac{3M}{2\pi R^3} \tag{5.6}$$

The principal normal stress difference is given by eqn (5.7),[26]

$$N_1 = \frac{2F}{\pi R^2} \tag{5.7}$$

where $F$ is the vertical thrust pushing apart the cone and plate.

Various parallel plate (disk) and enclosed disk (Mooney) rheometers (Fig. 5.1(c)) have been used on polymer melts and especially elastomers through the years.[5, 31-34] The shear rate at the outer radius of the disk in a simple disk viscometer is

$$\dot{\gamma} = \frac{R\Omega}{H} \tag{5.8}$$

where $R$ is the disk radius and $H$ the disk separation. The shear stress at the outer radius of the disk for a simple parallel disc viscometer is:

$$\sigma_{12}(R) = \left(\frac{N+3}{4}\right)\frac{M}{2\pi R^3} \tag{5.9a}$$

where

$$N = \frac{d \log (M/2\pi R^3)}{d \log (R\Omega/H)} \tag{5.9b}$$

and $M$ is the torque required to keep one disk stationary. For an enclosed disk viscometer, the shear stress at the outer radius is

$$\sigma_{12}(R) = \left(\frac{N+3}{4}\right)\frac{M}{2\pi R^3}F(b/R) \tag{5.10}$$

where $R$ is the outer radius of the disk and $b$ the distance between the outer radius of the disk and the surrounding cavity. The quantity $F(b/R)$ may be determined by analyzing the fluid mechanics.[34]

The most widely used rheological instrument to study shear flow of filled polymer melts is the capillary rheometer (Fig. 5.1(d)).[6, 16, 20-25, 28, 35-39] The shear rate at the capillary die wall is[30]

$$\dot{\gamma}_w = \left(\frac{3n' + 1}{4n'}\right)\frac{32Q}{\pi D^3} \tag{5.11}$$

where

$$n' = \frac{d \log (\sigma_{12})_w}{d \log 32Q/\pi D^3} \tag{5.12}$$

The die wall shear stress $(\sigma_{12})_w$ is determined from Bagley plots of total pressure $p_t$ versus die length $(L)$/diameter $(D)$ ratio:

$$p_t = 4(\sigma_{12})_w\frac{L}{D} + \Delta p_{ends} \tag{5.13}$$

Here $\Delta p_{ends}$ is the ends pressure loss.

Slit rheometers and their applications have also been described in the literature. This instrument has been particularly applied by Han and his coworkers.[40-42] The shear rate at the die wall is

$$\dot{\gamma}_w = \left(\frac{2n'' + 1}{3n''}\right)\frac{6Q}{WH^2} \tag{5.14}$$

where $H$ is the slit thickness and $W$ is the width. Here

$$n'' = \frac{d \log (\sigma_{12})_w}{d \log 6Q/WH^2} \tag{5.15}$$

Exit pressure losses in slit rheometers have been used to compute normal stresses.

### 5.3.2. Elongational Flows

Elongational flows involve velocity fields of form:

$$\mathbf{v} = v_1(x_1)\mathbf{e}_1 + v_2(x_2)\mathbf{e}_2 + v_3(x_3)\mathbf{e}_3 \tag{5.16}$$

The velocity gradients are in the direction of flow rather than orthogonal to it.

Most studies of elongational flow have involved uniaxial extension where

$$v_2(x_2) = v_3(x_3) \tag{5.17}$$

and in an instrument where a constant stretch rate is prescribed,

$$v = Ex_1 e_1 + \left(-\frac{E}{2}\right)x_2 e_2 + \left(-\frac{E}{2}\right)x_3 e_3 \qquad (5.18)$$

The stress response to this flow is uniaxial in character and may be represented by

$$\sigma_{11} = \chi E; \qquad \sigma_{22} = \sigma_{33} = 0 \qquad (5.19)$$

Here $\chi$ is a uniaxial elongational viscosity function. We may similarly define biaxial elongational viscosity functions $\chi_B$ based on equality of $v_1(x_1)$ and $v_2(x_2)$ as well as $\sigma_{11}$ and $\sigma_{22}$. A planar elongational viscosity function $\chi_p$ may be defined from taking $v_2(x_2)$ equal to zero.

A variety of instruments have been developed to measure the rheological functions $\chi, \chi_B$ and $\chi_p$. We shall describe only instruments to characterize $\chi$ here. These instruments may be divided according to the method: (i) variable sample length–constant stretch rate,[43,44] (ii) constant sample length–constant stretch rate,[29,45–47] and (iii) constant stress measurement[48–50] (Fig. 5.2).

In all of these instruments the sample, a filament, is immersed in or drawn on the top of a bath. In the variable sample length instrument a constant elongation rate is imposed and the sample length is varied in such a way that $E$ is constant. Thus

$$E = \frac{1}{L}\frac{dL}{dt} \qquad (5.20a)$$

$$L = L(0)\exp(Et) \qquad (5.20b)$$

In constant sample length–constant stretch rate instruments, the sample may be held between a clamp (LVDT) and a rotating wheel (or pair of wheels) which removes it from the bath. Here

$$E = \frac{V}{L} \qquad (5.21a)$$

where $V$ is the wheel velocity. In an instrument devised by Meissner[45] two sets of wheels are used, one of which senses the tension. The elongational rate is then

$$E = \frac{V}{L/2} \qquad (5.21b)$$

In the constant stress instruments, a cam is used to draw out the filament
and its motion as a function of time is recorded.

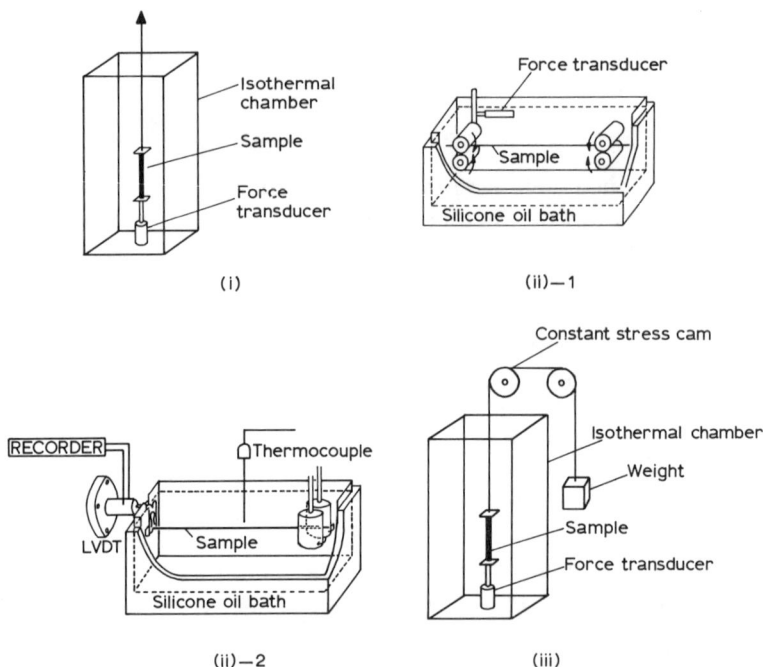

**Fig. 5.2.** Elongational flow rheometers: (i) variable sample length, constant
stretch rate; (ii) constant sample length, constant stretch rate (1, Meissner;
2, Minoshima–Suetsugu–Yamane); (iii) constant stress.

## 5.4. EXPERIMENTAL STUDIES OF RHEOLOGICAL
## PROPERTIES OF FILLED POLYMER MELTS

### 5.4.1. Observations for Concentrated Suspensions in
### Newtonian Fluids

There is a long history of investigations of rheological properties of
suspensions of small particles. Before proceeding to describe filled
polymer melts, it is useful to recall the studies of the rheological
properties of suspensions of small particles in Newtonian fluid
mechanics. These systems received extensive investigation in the 1920s
and 1930s. The key investigator was Herbert Freundlich, whose

observations are summarized in a series of papers which appeared in 1934–1938.[51-56] Freundlich and Jones[53] noted that suspensions generally exhibited three distinctive rheological characteristics in shear flow experiments, which they express as dilatancy, plasticity and thixotropy. Dilatancy implies that the suspensions dilated when sheared and generally exhibited an increasing viscosity with increasing shear rate.[56] Plasticity indicates the occurrence of yield values and implicitly a viscosity which decreases with shear rate. Thixotropy implies a viscosity which decreases with extent of shear. Individual suspensions obviously cannot exhibit all these characteristics. Freundlich sought to determine the structural features of concentrated suspensions and relate these to the above rheological characteristics. First consider roughly spherical particles. The primary variable found was the degree of aggregation of the particles as observed by optical microscopy and most simply measured by sedimentation volume, i.e. the observed volume of a sediment in a liquid relative to the volume computed from density measurements. Dilatancy was associated with small sedimentation volumes (small levels of aggregation). Plasticity and thixotropy were found with systems exhibiting large sedimentation volumes (high degrees of aggregation).

Freundlich and Jones[53] associated the sedimentation volume with particle size and interparticle forces. Large sedimentation volumes were found with small particle sizes: this behavior was exhibited for particle sizes of 1–10 $\mu$m and smaller. Repulsive forces between particles prevent aggregation and reduce sedimentation volume.

Freundlich[52, 55] also considered the influence of particle anisotropy. Concentrated suspensions of rodlike particles (e.g. vanadium pentoxide) were found to be birefringent when observed between crossed polars.[55] These structures were called *tactoids* and represent the necessary local particle alignments which must occur in concentrated suspensions of rods. Suspensions of anisotropic particles were invariably found to be thixotropic. This was even the case with systems which did not exhibit yield values.

Various additional striking observations have been made in the period since 1950, most notably with non-linear effects in chopped fiber suspensions. Nawab and Mason[57] and later investigators[58, 59] have shown that chopped fiber suspensions in Newtonian fluids exhibit strong Weissenberg effects and apparently large normal stresses. Mewis and Metzner[58] have found that chopped fiber suspensions in Newtonian fluids exhibit large elongational viscosities.

## 5.4.2. Shear Flow

*Pure Polymer Melts*
Before it is possible to discuss rationally the characteristics of particle
filled polymer melts, it is necessary to describe the behavior of the
matrix fluid, which is complex. It is generally agreed that polymer melts
are viscoelastic fluids, whose behavior may be represented by various
constitutive relationships which have been summarized in several
monographs.[40, 60, 61] The concern here, however, is with the behavior of
those materials in shear flows.

There have been very extensive investigations of the flow behavior of
polymer melts and the influence of molecular structure on those
characteristics. Reduced shear viscosity $\eta(\dot\gamma)/\eta_0$ is found to be
independent of temperature if plotted versus $\eta_0\dot\gamma$. Such plots are given in
Vinogradov and Malkin.[62] The principal normal stress difference $N_1$ is
found to be independent of temperature if plotted as a function of $\sigma_{12}$.[63]
Both plots are, however, dependent upon molecular weight distribution

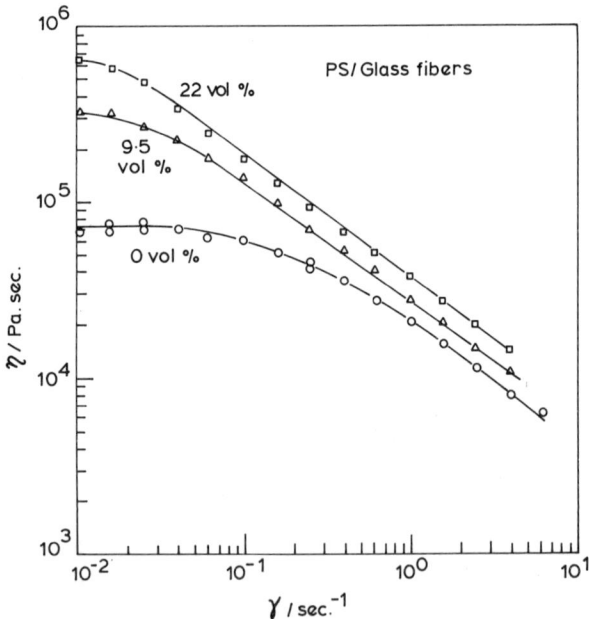

**Fig. 5.3.** Shear viscosity of polystyrene–chopped glass fiber suspension as a
function of a shear rate at 180 °C.

in linear polymer systems and upon long chain branching. Broadening of the molecular weight distribution causes (i) $\eta/\eta_0$ to fall off more rapidly with $\eta_0\dot\gamma$,[47, 64] and (ii) $N_1$ to increase at constant $\sigma_{12}$.[64, 65]

## Filled Polymer Melts

The observations of rheological properties made on particle filled polymer melts and elastomers generally correspond to those made on suspension of particles in Newtonian fluid media. Systems containing large glass spheres and fibers exhibit low shear rate Newtonian viscosities.[22, 26, 27, 60] At higher shear rates, the viscosity decreases in a manner similar to pure polymer melts (Fig. 5.3). Czarnecki and White[26] have found that plots of $\eta/\eta_0$ versus $\eta_0\dot\gamma$ for polymers filled with large particles are independent of fiber loading. This appears to be the case for most particles whose smallest dimension is 10 $\mu$m or more.

Particles with submicron dimensions generally produce different effects on the rheological properties of suspensions in polymer melts. Concentrated suspensions of such particles generally exhibit thixotropy and yield values in shear flows. We summarize experimental studies of compounds exhibiting yield values in Table 5.1. It appears that yield

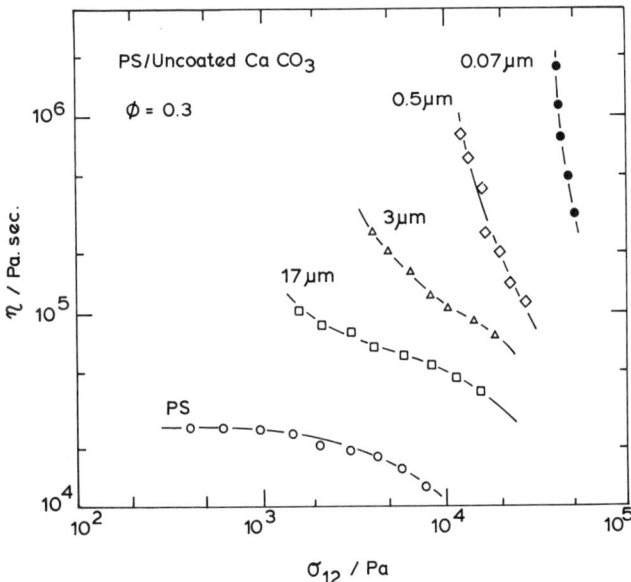

Fig. 5.4. Shear viscosity of polystyrene–calcium carbonate compounds as a function of shear stress with different particle sizes; $\phi = 0.3$.

TABLE 5.1

Investigations of Shear Viscous Behavior of Particle Filled Polymer Melts/Elastomers Showing Yield Values

| Particles | Particle dimensions ($\mu m$) | Polymer matrix | Instrument | Ref. | Comment |
|---|---|---|---|---|---|
| Carbon black | 0·1 | | Parallel plate | Zakharenko et al.[13] | Wide range of |
| | 0·02 | Polyisobutylene | Parallel plate | Vinogradov et al.[19] | experiments. |
| | | | Cone–plate | | |
| | | | Coaxial cylinder | | |
| | | | Capillary | | |
| | 0·025 | Polystyrene | Cone–plate | Lobe and White[24] | Normal stresses measured. |
| | | | Capillary | | |
| | 0·045 | Polystyrene | Cone–plate | Tanaka and White[25] | Normal stresses measured. |
| | 0·03 | Butadiene–styrene copolymer | Sandwich | Toki and White[16] | |
| | | | Capillary | | |
| | 0·03 | Natural and synthetic polyisoprenes | Sandwich | Montes and White[67] | |
| | | | Capillary | | |

| | | | | | |
|---|---|---|---|---|---|
| Calcium carbonate | 0·5 | Polyethylene polystyrene | Cone–plate | Kataoka et al.[22,23] | Normal stresses measured. Surface coating effect. |
| | | Polystyrene | Cone–plate | Tanaka and White[25] | Normal stresses measured. Surface coating effect. |
| | 0·07, 0·5, 3, 17 | Polystyrene | Cone–plate | Suetsugu and White[29] | Normal stresses measured. Surface coating effect. |
| Talc | | Polypropylene | Cone–plate Capillary | Chapman and Lee[18] | Surface coating effect |
| Titanium dioxide | 0·18 | Polyethylene | Cone–plate Capillary | Minagawa and White[20] | Normal stresses measured. Surface coating effect. |
| | 0·18 | Polystyrene | Cone–plate | Tanaka and White[25] | Normal stresses measured. |
| Franklin fiber | 0·5–2 | Polystyrene | Cone–plate | White et al.[27] | Normal stresses measured. |

values begin to occur at volume loadings of about 0·15,[16, 18–20,24, 25, 27, 29] and increase with increasing volume fraction. Generally, the magnitude of the shear viscosity and the yield value increase with decreasing particle size at fixed volume fraction (Fig. 5.4). Suetsugu and White[29] found the yield value increases roughly inversely with the particle diameter.

Surface coatings on particles can cause substantial reductions in viscosity.[18, 21, 25, 29] This has been observed for several particle coated systems including talc,[18] titanium dioxide[20] and calcium carbonate.[25, 29, 42] A case in point is calcium carbonate coated with stearic acid (Fig. 5.5).

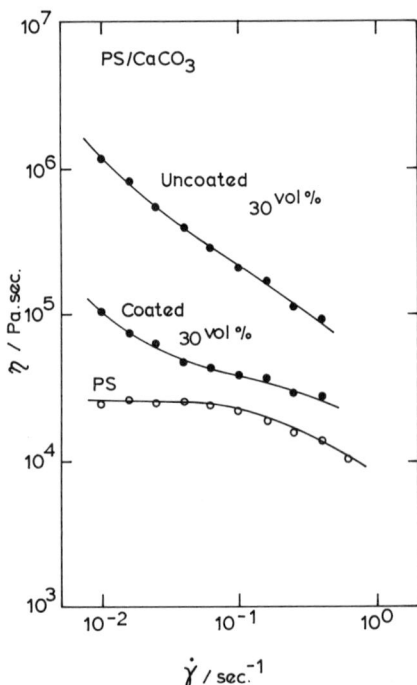

**Fig. 5.5.** Viscosity–shear rate data for coated and uncoated calcium carbonate filled polystyrene.

The magnitudes of viscosity increase and yield value seem to correlate with magnitude of interparticle forces, being greater for particle systems with polar structures (e.g. talc, calcium carbonate) as opposed to non-polar particles (e.g. carbon black) where only van der Waals bonding is involved.

I now turn to normal stresses. Chan *et al.*[21] have found that normal stress difference $N_1$ is greatly enhanced by the presence of chopped glass fibers. This has been studied more extensively by Czarnecki and White.[26] They have investigated $N_1$ in suspensions of aramid and cellulose as well as of glass fibers. It is found that the dependence of $N_1$ upon $\sigma_{12}$ increased with both the rigidity and aspect ratio of the fibers. Mastication of suspensions of chopped glass fibers not only damaged the fibers and decreased their aspect ratios, but lowered $N_1$. The value of $N_1$ eventually decreased to the correlation obtained for the $N_1 - \sigma_{12}$ response of the melt matrix (Fig. 5.6).

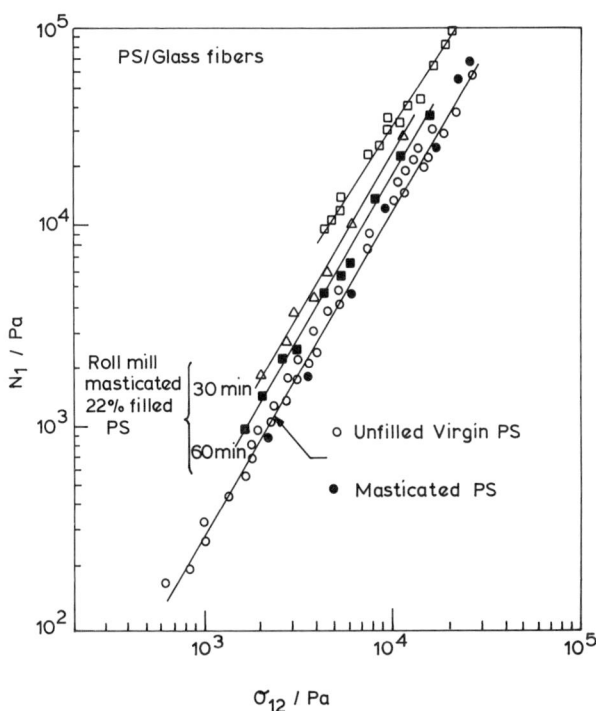

**Fig. 5.6.** Influence of chopped fibers on normal stress–shear stress relationship for polystyrene.

The problem of normal stress measurement in polymer melts filled with small particles has been discussed in some detail by Suetsugu and White.[29] In cone-plate (or parallel disk) total thrust measurement instruments, the stresses developed during sample preparation do not completely decay away. This is associated with these materials exhibiting yield values. This makes interpretation of the results difficult. Where measurements are possible, it is generally found that normal stresses are reduced at specified shear stress.[24, 25, 27, 29, 41]

### 5.4.3. Elongational Flow

*Pure Polymer Melts*
Studies of uniaxial elongational flow of polymer melts date back a generation. Problems exist in many polymer melt systems in being sure that uniform filaments are formed. Generally low-density polyethylene, polystyrene and moderately narrow distribution polyolefins may be drawn out into uniform filaments.[46, 47, 64] Broad molecular weight distribution polymers such as typical polyethylene, polypropylene and polybutene-1 readily develop necks and fail at short extension ratios. The following comments focus on those melts where clearly stable uniform filaments are formed.

At low stretch rates, the uniaxial elongational viscosity of those systems achieves a stable value of three times the shear viscosity, $3\eta_0$. At

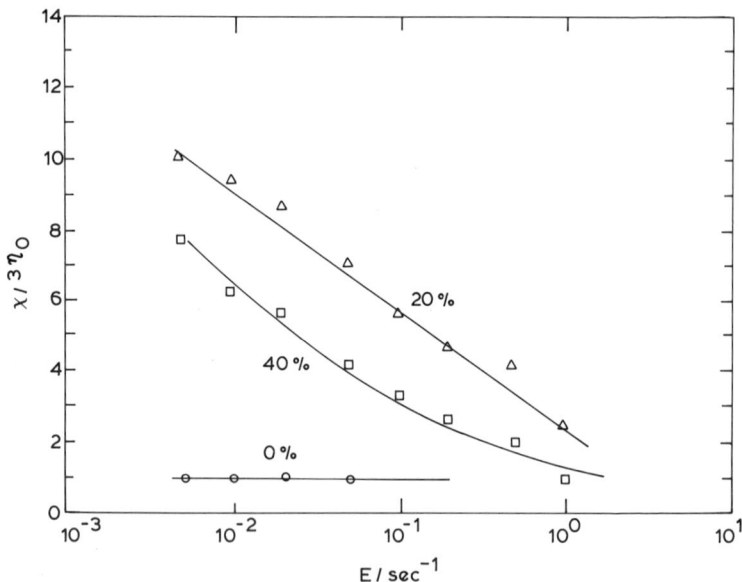

**Fig. 5.7.** Influence of chopped fibers on reduced elongation viscosity, $\psi/3\eta_0$.

higher stretch rates the elongational viscosity increases.[45-47, 64, 68] This is all that is found in some instruments. In other instruments, notably that developed by Meissner[45] and upgraded by Laun and Munstedt,[68] a maximum is observed followed by a decreasing elongational viscosity function.

*Filled Polymer Melts*
The elongational viscosity has been reported in particle filled polymer melts by the present author and his coworkers.[21, 24, 25, 29] Chan *et al.*[21] found that polymers filled with chopped fibers contain greatly enhanced elongational viscosities (Fig. 5.7). These viscosity functions decrease with increasing stretch rates in a manner similar to the shear viscosity function. The 20 wt % material has a higher $\chi/3\eta_0$ than the

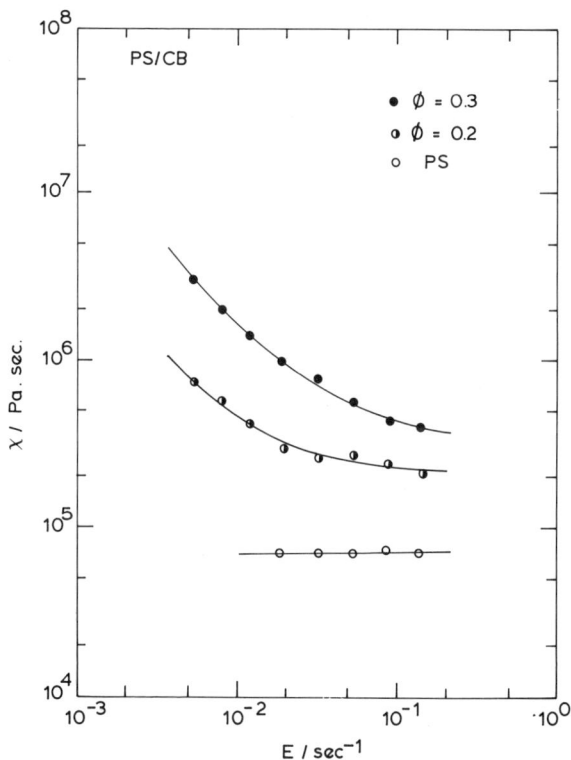

Fig. 5.8. Elongational viscosity of carbon black compounds of polystyrene.

40 wt % material. This is presumably because the latter fibers have shorter aspect ratios.

We now turn to suspensions of small particles in polymer melts. The elongational viscosity behavior of carbon black-polystyrene compounds has been examined by Lobe and White[24] and Tanaka and White.[25] The viscosity increased indefinitely at low stretch rates and a yield value was observed (Fig. 5.8). Tanaka and White[25] and Suetsugu and White[29] observed yield values in calcium carbonate-polystyrene suspensions in elongational flow. The former authors also observed yield values in the titanium dioxide-polystyrene system.[25] All of the systems exhibiting yield values in elongational flow also exhibit this behavior in shear.

## 5.5. THEORIES OF FLOW OF PARTICLE FILLED SYSTEMS

### 5.5.1. Mechanistic Theories of Viscosity

There is a long history of mechanistic investigations of the influence of particulates on the rheological properties of suspensions. Almost all of these studies involve infinitely dilute systems and it has only been since 1950 that studies of concentrated suspensions have appeared. The classical study which initiated research in this area is that of Einstein,[69] who showed that the presence of rigid spheres increased the viscosity of a Newtonian fluid according to the expression

$$\eta = \eta_0 [1 + 2 \cdot 5\phi] \tag{5.22}$$

where $\eta_0$ is the viscosity of the matrix and $\phi$ the volume fraction of particles. Einstein's theory was generalized by Jeffery[70] to the case of ellipsoidal particles. Jeffery found for infinitely dilute suspensions of ellipsoids that

$$\eta = \eta_0 [1 + C\phi] \tag{5.23}$$

where $C$ depends upon particle aspect ratio and initial orientation. Batchelor[71, 72] presented a significant generalization of research in this area by computing stress fields in dilute suspensions and showing how the results could lead to complete constitutive equations (see Section 5.5.2).

It is implicit in Einstein's work that the elongational viscosity of a dilute suspension of spheres will be three times the shear viscosity. This is made explicit by Batchelor.[71] This will not be the case for anisotropic particles. This has also been explored in detail by Batchelor.[72] If, as is likely, the anisotropic particles are oriented with their major axis in the direction of flow, the elongational viscosity is predicted to be much larger than the shear viscosity.

Studies of the influence of concentration on the viscosity of suspensions date to Guth and Simha,[73] who made perturbation studies of the interaction in steady viscous flow. Various later investigations have considered this problem, the most recent being Batchelor and Green,[74] who concluded that, to second-order terms, the viscosity increases according to

$$\eta = \eta_0 [1 + 2{\cdot}5\phi + 7{\cdot}6\phi^2] \tag{5.24}$$

Highly concentrated suspensions of non-interacting spheres have been considered by Simha[75] and later investigators using hydrodynamic lubrication theory.[76, 77] These papers neglect aggregation effects and consider the spheres to be essentially arrayed in a lattice. Frankel and Acrivos[77] develop the simple form

$$\frac{\eta}{\eta_0} = \frac{9}{8}\left[\frac{(\phi/\phi_m)^{1/3}}{1 - (\phi/\phi_m)^{1/3}}\right] \tag{5.25}$$

There have been many studies of the viscosity of suspensions of spheres with electrical charges or with strong van der Waals forces between them.[78-85] Attractive forces generally cause aggregation and increase viscosity.

Concentrated suspensions of fibers have been treated by Batchelor,[72] who has modelled their response in elongational flow. He argues that the flow between the parallel fibers will be shearing in character and lubrication theory may be applied. In particular Batchelor predicts

$$\chi = 3\eta_0\left[1 + \frac{4}{9}\left(\frac{L}{D}\right)^2 \frac{\phi}{\ln \pi/\phi}\right] \tag{5.26}$$

where $L$ and $D$ are the length and diameter of the fibers. This expression has been verified by Mewis and Metzner.[58] Fiber aspect ratio plays a strong role.

There have been few investigations of the influence of particles on the rheological properties of non-Newtonian fluid media. Goddard[86] has developed a theory of elongational viscosity of suspensions of fibers in a non-Newtonian fluid matrix. Tanaka and White[85] have generalized the theory of Frankel and Acrivos[77] to include not only interparticle interaction but a non-Newtonian fluid matrix. Particle aggregation is neglected.

## 5.5.2. Constitutive Equations

*Pure Polymer Melts*
It is generally agreed that flexible chain polymer melts and elastomers exhibit viscoelastic behavior, stresses being a hereditary functional of the history of deformation. Linear one-dimensional behavior is representable in alternative forms as (i) a differential equation in stress $\sigma$ and strain $\gamma$, specifically

$$a_0\sigma + \sum_{n=1}^{n=N} a_n \frac{\mathrm{d}^n\sigma}{\mathrm{d}t^n} = \sum_{m=1}^{m=M} b_m \frac{\mathrm{d}^m\gamma}{\mathrm{d}t^m} \tag{5.27}$$

or (ii) with $\sigma$ as an integral over the history of $\mathrm{d}\gamma/\mathrm{d}t$ or $\gamma$

$$\sigma = \int_{-\infty}^{t} G(t-s)\frac{\mathrm{d}\gamma}{\mathrm{d}s}\,\mathrm{d}s = \int_{-\infty}^{t} \Phi(t-s)\,\gamma(s)\,\mathrm{d}s \tag{5.28}$$

Equation (5.28) may be obtained from eqn (5.27) from the method of Laplace transforms. For a Maxwell model:

$$a_0 = -1/\tau; \quad a_1 = 1; \quad a_j = 0 \quad (j > 2) \tag{5.29a}$$

$$b_1 = G; \quad b_j = 0 \quad (j > 1) \tag{5.29b}$$

$$G(t) = G\exp(-t/\tau) \quad \Phi(t) = \frac{G}{\tau}\exp(-t/\tau) \tag{5.30}$$

Non-linear three-dimensional formulations of viscoelasticity date to the work of Zaremba[87] in 1903. The modern formulation of the theory is associated with the work of Oldroyd,[88, 89] Lodge,[90, 91] Rivlin and his coworkers[92, 93] and Coleman and Noll.[94, 95] This research has been summarized in various monographs.[30, 40, 60, 61] The differential constitutive equation generalizing eqn (5.27) is of the form:

$$\sigma = -p\,\mathbf{I} + \mathbf{T} \tag{5.31}$$

$$a_0\mathbf{T} + \sum_{n=1}^{} a_n \frac{\delta^n\mathbf{T}}{\delta t^n} = b_1\mathbf{d} + \sum_{m=1}^{} b_{m+1} \frac{\delta^m}{\delta t^m}\mathbf{d} + c_1\mathbf{d}^2 \tag{5.32}$$

$$+ c_2(\mathbf{T}\cdot\mathbf{d} + \mathbf{d}\cdot\mathbf{T}) + ....$$

where the $c_j$ terms represent new tensor components involving cross-products of $T$, $d$ and their derivatives. The $\delta/\delta t$ are Oldroyd *convected derivatives* which may be expressed

$$\frac{\delta T}{\delta t} = \frac{\partial T}{\partial t} + (v \cdot \nabla)T - T \cdot \nabla v - \nabla v \cdot T \qquad (5.33)$$

The integral constitutive equations, generalizing eqn (5.28), usually are expressed as:

$$T = \int_{-\infty}^{t} [m_1(t-s)\,c^{-1} - m_2(t-s)\,c]\,ds \qquad (5.34)$$

where $m_1(t)$ and $m_2(t)$ are relaxation modulus functions which may depend upon variants of $T$, $d$, $c^{-1}$ and $c$. The quantities $c^{-1}$ and $c$ are the Finger and Cauchy deformation tensors.

*Chopped Fiber Suspensions*
Mechanistic formulations of flow behavior of suspensions of aniso-tropic particles leading to tensor constitutive equations derive from Batchelor.[71, 72] Evans[96] has described the application of this theory to solve hydrodynamic problems of varying levels of complexity. They use a dilute solution formulation (first order in $\phi$). This approach has been expanded in papers by Hinch and Leal[97] and Bark and Tinoco.[98] The constitutive equations discussed have the general form:

$$T = 2\eta_0 d + 4\eta_0\phi\,[A\,\overline{pppp}{:}\,d + B(\overline{pp}\cdot d + d\cdot\overline{pp}) + Cd + D\overline{pp}] \qquad (5.35)$$

Here $A$, $B$, $C$ and $D$ are shape-dependent factors and $p$ is a unit vector representing the orientation of the major axis. The $\overline{pp}$ and $\overline{pppp}$ represent second and fourth moments of $p$ with the probability distribution. For spheres the terms in $A$, $B$ and $D$ are zero. For rigid fibers we need only consider the term in $A$.

*Small Particle Polymer Melt Suspensions*
Phenomenological theories of the flow of suspensions are based on representations of the responses of materials exhibiting yield stresses, $Y$. One-dimensional formulations date to Schwedoff,[99] Bingham[100, 101] and Buckingham.[102] A material behaving in a differentially viscous manner above the yield value was represented as

$$\sigma = Y + \eta_B\dot{\gamma} \qquad (5.36)$$

This is known as a Bingham plastic. Alternative and more complex formulations which allow for non-Newtonian differential response were proposed by Herschel and Bulkley[103] in the form

$$\sigma = Y + k\dot{\gamma}^n \tag{5.37}$$

and by Casson[104] as

$$\sigma^{1/2} = Y^{1/2} + k\dot{\gamma}^{1/2} \tag{5.38}$$

Viscoelastic response above the yield value was represented by Schwedoff[99] in terms of a modified Maxwell model (compare eqns (5.27) and (5.29)).

$$\frac{d\sigma}{dt} = G\frac{d\gamma}{dt} - \frac{1}{\tau}(\sigma - Y) \tag{5.39}$$

Three-dimensional constitutive formulations of this behavior for viscous fluid response above the yield value date to Hohenemser and Prager[105] and Oldroyd.[106] They used the von Mises formulation of the yield surface

$$\text{tr } \mathbf{P}^2 = 2Y^2 \tag{5.40}$$

where $\mathbf{P}$ is the deviatoric stress tensor defined by

$$\sigma = 1/3 \, (\text{tr } \sigma) \, \mathbf{I} + \mathbf{P} \tag{5.41}$$

For a material which exhibits differentially viscous response above the yield value (three-dimensional Bingham plastic) we have

$$\mathbf{P} = \left[ \frac{Y}{\sqrt{\frac{1}{2} \text{tr } \mathbf{P}^2}} \right] \mathbf{P} + 2\eta_B \mathbf{d} \tag{5.42}$$

or as shown by Oldroyd[106]

$$\mathbf{P} = \left[ \frac{2Y}{\sqrt{2 \, \text{tr } \mathbf{d}^2}} \right] \mathbf{d} + 2\eta_B \mathbf{d} \tag{5.43}$$

Thixotropy was introduced into this formulation by Slibar and Pasley.[107] They represented $Y$ as a hereditary functional of $\text{tr } \mathbf{d}^2$ to accomplish this.

Oldroyd[108] has generalized eqn (5.43) so that it may apply to materials which are non-Newtonian above the yield value.

The three-dimensional theory of materials which are differentially viscoelastic above the yield value has been developed by Hutton[109] and more explicitly by White.[110-112] This takes the form:[110]

$$P = \left[ \frac{Y}{\sqrt{\frac{1}{2} \text{tr } P^2}} \right] P + H \tag{5.44a}$$

or equivalently,[110]

$$P = \left[ \frac{Y}{\sqrt{\frac{1}{2} \text{tr } H^2}} \right] H + H \tag{5.44b}$$

Here $H$ is a hereditary functional constitutive equation similar to eqn (5.34) but deviatoric in character. A possible form might be

$$H = \int_{-\infty}^{t} [m_1(t-s)(c^{-1} - \tfrac{1}{3}(\text{tr } c^{-1}) I)$$

$$- m_2(t-s)(c - \tfrac{1}{3}(\text{tr } c) I)] \, ds \tag{5.45}$$

Thixotropic effects have been introduced into this formulation by Suetsugu and White.[113] They develop an expression for the yield function $Y$ making it a hereditary functional of $\text{tr } d^2$. The plastic viscoelastic fluid of eqn (5.44), with constant $Y$, is a long-duration steady-state asymptote of their theory.

Comparisons of the theory described in this section with experimental data on compounds of calcium carbonate, carbon black and titanium dioxide in polystyrene are described in papers by White and Tanaka[112] and Suetsugu and White.[113]

## 5.6. PROCESSING PARTICLE FILLED MELTS

### 5.6.1. Mixing

Fibers and small particles may be mixed into polymer melts using a range of commercial mixing apparatus including mills, internal (Banbury) mixers, single screw and twin screw extruders. The classic paper on dispersion of particulates in mills and internal mixers is the review of Bergen.[114] The subject of compounding extruders has recently been reviewed by Hermann.[115]

Generally, mixing is dominated by shear stress and work. High shear stresses and high power produce good dispersion. Mixing at low temperatures where the viscosity is large is usually effective. Scale-up often results in problems because large systems cannot lose the heat built up by viscous dissipation and they operate at higher temperatures. One hopes in mixing that the stresses developed will break up aggregates of particles. However, if the suspended particles are brittle fibers, they will usually be damaged. This problem of fiber breakage during dispersion processes has been explored by O'Connor[116] and by Czarnecki and White.[26] Generally, cellulose and synthetic organic fibers are more resistant to breakage than glass or carbon fibers.

## 5.6.2. Extrusion

*Chopped Fiber Suspensions*
Studies of the influence of particulates on the extrusion characteristics of polymers have several aspects. These include flow visualization, pressure losses in dies and the characteristics of emerging extrudates.

Flow visualization of a suspension of chopped glass fibers in a Newtonian (epoxy) matrix has been studied by Murty and Modlen,[117] Lee and George[118] and Crowson *et al.*[119] Lee and George observed a plug flow orientation in the reservoir which transformed towards a parabolic profile as it converged into a die. Significant fiber orientation during the converging flow was observed. Streamline flow was found in convergent die entries, but vortices were noted in 180° entrance dies. Murty and Modlen,[117] using glass fibers in glycerin, similarly found that fibers orient in converging flow. They also observed a jamming effect in concentrated solutions. Crowson *et al.*[119] found similar flow alignment in the converging entrance flow region. They noted that shear flow in the die produced decreases in orientation.

Chan *et al.*[120] found that fiber suspensions in polymer melts exhibit very large ends pressure losses (i.e. die entrance plus exit losses) when compared with homogeneous polymer melts. These may be associated with fiber alignment, jamming and the large elongational viscosities exhibited by fiber suspensions.[121]

The swell and character of chopped fiber filled melts has been described by various researchers.[28, 120, 122] Generally, swell is significantly reduced by the presence of chopped fibers (Fig. 5.9).

The orientation of fibers in extrudates has been investigated by Cessna[123] and various later investigators.[28, 124-127] Most studies of orientation have been visual and qualitative with observers concluding that fiber orientation relative to the die axis increases with extrusion rate. Menendez and White[128] used wide-angle diffraction with aramid

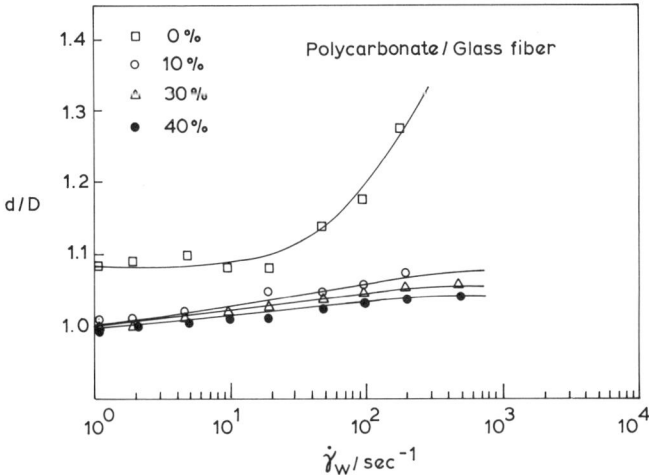

**Fig. 5.9.** Extrudate swell of polymer melts containing chopped fibers.

chopped fibers containing high levels of crystalline orientation to obtain quantitative measurements of fiber orientation in extrudates. Specifically, they computed the Hermans' orientation factor:

$$f = \frac{3 \cos^2 \phi - 1}{2} \tag{5.46}$$

where $\phi$ is the angle between the direction of extrusion and the fiber axis. This has values of order 0·15 in the reservoir preceding the die entry region, 0·47 within the capillary and 0·30 in the extrudate.

Orientation of fibers in the direction of flow is considered undesirable in tubing and hose. Cessna[123] has proposed the use of dies with rotating members to give helical orientations of fibers in extrudates. Goettler *et al.*[124–126] have achieved the same goal using annular dies with diverging sections. The biaxial elongational flow induced by diverging sections introduces torques which rotate fibers into the direction of flow.

Cole and coworkers[129] found that when certain molten thermoplastics containing glass fibers up to 20 mm in length are extruded through an orifice, the extrudate has a frothy texture and solidifies into an irregular 'open cell' structure. This effect is apparently due to recovery of the fibers from distortions arising in the region of converging flow near the entry to the die. This was achieved with polypropylene containing 0·25 wt % fibers. It is most pronounced in systems with viscosity in the range 50–500 Pa s. Cole *et al.* call these extrudates 'fiber foam'.

*Small Particle Polymer Melt Suspensions*
Similar studies of the influence of small particulates on the extrusion of polymers have appeared. Ma *et al.,* [130] applying a technique of Ballenger and White,[131] have introduced flow markers into pure polymers and compounds in the reservoir preceding the die. Low-density polyethylene was the polymer matrix, with a carbon black filler. The pure polymer is well known to exhibit striking large vortices in the die entry region.[132, 133] The marked reservoir material was cooled down following flow, removed from the apparatus and sectioned to analyze the streamlines. It was found that the pure low-density polyethylene exhibited the well-known vortices as expected. However, the addition of more than 5 vol. % of carbon black to the polymer was found to eliminate the vortices and induce what appears to be a uniform streamline flow similar to that classically observed in high-density polyethylenes.

There have been few studies concentrating on die ends pressure losses in filled thermoplastics. White and Crowder[39] observed that the addition of carbon black to elastomers reduces the ratio of ends pressure losses to capillary wall shear stress.

There is a long history of investigations of the influence of particulates on the characteristics of extrudates. Papers by McCabe and Mueller,[35] Hopper,[37] Collins and Oetzel,[38] Vinogradov *et al.,*[19] White and Crowder,[39] Minagawa and White[20] and Lobe and White[24] described the influence of a range of particulates including carbon black, titanium dioxide and clay on extrudate swell. It is found that the addition of particulates reduces extrudate swell and increases the range of operating conditions where smooth extrudates are observed (Fig. 5.10).

White and Huang[134] have considered the hydrodynamic mechanism of the reduction of extrudate swell due to the addition of particulates. They associate it with the existence of yield values in the compounds.

### 5.6.3. Injection Molding
Studies of injection molding suspensions of chopped fibers in polymer melts have involved mold filling and fiber orientation. Oda *et al.,*[135] by filling end-gated molds, have clearly shown the existence of two regimes. The first regime is simple mold filling where a front filling the

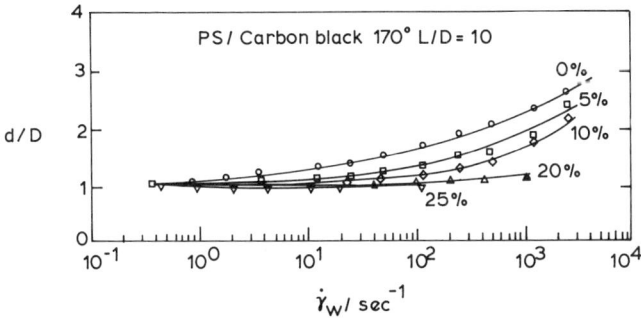

**Fig. 5.10.** Extrudate swell of carbon black filled polystyrene.

mold at that position expands away from the gate into the mold. The second is a jet which shoots into the mold from the gate. They show that the transition between the regimes is determined by the magnitude of the extrudate swell $d$ relative to the thickness $H$ of the mold. As we have noted, glass fibers significantly depress the swell of extrudates. It should not be surprising that Chan *et al.*[120] have found, by flow visualization with transparent molds, that jetting is a serious problem with end-gated molds for fiber filled thermoplastics.

The development of orientation of chopped fibers in molded parts was first extensively studied by Goettler.[136, 137] Later studies have been reported by Darlington and McGinley,[138] Owen and his coworkers,[139, 140] Crowson *et al.*[119, 122] and Oyanagi *et al.*[141] As first shown by Goettler and confirmed by later investigators, whilst converging flow orients fibers in the flow direction, diverging flow orients them in the transverse direction. The influences of injection rate and gate size have been discussed.

There have been few studies of the injection molding of small particle–polymer melt suspensions. Jetting again appears to be a problem in mold filling because of the small magnitude of swell.

## ACKNOWLEDGEMENT

I thank Mr Y. Suetsugu for his aid with the figures used in this chapter.

# REFERENCES

1. Couette, M., *Compt. Rend.,* 1888, **107**, 388.
2. Maxwell, J. C., *Phil. Trans. Roy. Soc.,* 1866, **156**.
3. Reynolds, O., *Phil. Trans. Roy. Soc.,* 1886, **177**, 157.
4. Mooney, M., *Physics,* 1936, **7**, 413.
5. Mooney, M., *Ind. Eng. Chem., Anal. Ed.,* 1934, **6**, 147.
6. Dillon, J. H. and Johnston, N., *Physics,* 1933, **4**, 225.
7. Eisenschitz, R., Rabinowitsch,, B. and Weissenberg, K., *Mitt. Deut. Materialpruf,* 1929, **9**, 91.
8. Mooney, M., *J. Rheology,* 1931, **2**, 210.
9. Weissenberg, K., *Nature,* 1947, **159**, 310.
10. Weissenberg, K., *Proc. 1st Int. Rheol. Cong.,* 1949, II.
11. Pollet, W. F. O. and Cross, A. H., *J. Sci. Inst.,* 1950, **27**, 209.
12. King, R. G., *Rheol. Acta,* 1966, **5**, 35.
13. Zakharenko, N. V., Tolstukhina, F. S. and Bartenev, G. V., *Rubber Chem. Technol.,* 1962, **35**, 326.
14. Middleman, S., *Trans. Soc. Rheol.,* 1969, **13**, 123.
15. Furuta, I., Lobe, V. M. and White, J. L., *J. Non-Newt. Fluid Mech.,* 1976, **1**, 207.
16. Toki, S. and White, J. L., *J. Appl. Polym. Sci.,* 1982, **27**, 3171.
17. Mullins, L. and Whorlow, R. H., *Trans. IRI,* 1951, **27**, 55.
18. Chapman, F. M. and Lee, T. S., *SPE J.,* 1970, **26**(1) 37.
19. Vinogradov, G. V., Malkin, A. Ya., Plotnikova, E. P., Sabsai, O. Y. and Nikolayeva, N. E., *Int. J. Polym. Mat.,* 1972, **2**, 1.
20. Minagawa, N. and White, J. L., *J. Appl. Polym. Sci.,* 1976, **20**, 501.
21. Chan, Y., White, J. L. and Oyanagi, Y., *Trans. Soc. Rheology,* 1978, **22**, 507.
22. Kataoka, T., Kitano, T., Sasahara, M. and Nishijima, K., *Rheol. Acta,* 1978, **17**, 140.
23. Kataoka, T., Kitano, T., Oyanagi, Y. and Sasahara, M., *Rheol. Acta,* 1979, **18**, 635.
24. Lobe, V. M. and White, J. L., *Polym. Eng. Sci.,* 1979, **19**, 617.
25. Tanaka, H. and White, J. L., *Polym. Eng. Sci.,* 1980, **20**, 449.
26. Czarnecki, L. and White, J. L., *J. Appl. Polym. Sci.,* 1980, **35**, 1217.
27. White, J. L., Czarnecki, L. and Tanaka, H., *Rubber Chem. Technol.,* 1980, **53**, 823.
28. Knutssen, B. A., White, J. L. and Abbas, K. B., *J. Appl. Polym. Sci.,* 1981, **26**, 2347.
29. Suetsugu, Y. and White, J. L., *J. Appl. Polym. Sci.,* 1983, **28**, 1481.
30. Middleman, S., *The Flow of High Polymers,* Wiley, New York, 1968.
31. Mooney, M., in *Rheology,* Vol. 2, F. R. Eirich (Ed.), Academic Press, New York, 1958.
32. Mooney, M., *Proc. Int. Rubber Conf.,* 1959.
33. Sakamoto, K., Ishida, N. and Fukusawa, Y., *J. Polym. Sci., A-2,* 1968, **6**, 1999.
34. Nakajima, N. and Harrel, E., *Rubber Chem. Technol.,* 1979, **52**, 962.
35. McCabe, C. C. and Mueller, N., *Trans. Soc. Rheology,* 1961, **5**, 329.

36. Smit, P. P. A., *Rheol. Acta,* 1969, **8**, 277.
37. Hopper, J. R., *Rubber Chem. Technol.,* 1967, **40**, 463.
38. Collins, E. A. and Oetzel, J. T., *Rubber Age,* 1970, **102**, 64.
39. White, J. L. and Crowder, J. W., *J. Appl. Polym. Sci.,* 1974, **18**, 1013.
40. Han, C. D., *Rheology in Polymer Processing,* Academic Press, New York, 1976.
41. Han, C. D., *J. Appl. Polym. Sci.,* 1974, **18**, 821.
42. Han, C. D., Sandford, C. and Yoo, H. J., *Polym. Eng. Sci.,* 1978, **18**, 849; Han, C. D., van den Weghe, T., Shete, P. and Haw, J. R., *Polym. Eng. Sci.,* 1981, **21**, 196.
43. Ballman, R. L., *Rheol. Acta,* 1965, **4**, 137.
44. Vinogradov, G. V., Radushkevich, B. V. and Fikham, V. D., *J. Polym. Sci.,* *A-2,* 1970, **8**, 1.
45. Meissner, J., *Rheol. Acta,* 1969, **8**, 78; 1971, **10**, 230.
46. Ide, Y. and White, J. L., *J. Appl. Polym. Sci.,* 1978, **22**, 1061.
47. Yamane, H. and White, J. L., *Polym. Eng. Rev.,* 1982, **2**, 167.
48. Cogswell, F. N., *Plastics Polym.,* 1968, **36**, 109.
49. Vinogradov, G. V., Fikham, V. D. and Radushkevich, B. V., *Rheol. Acta,* 1972, **11**, 286.
50. Munstedt, H., *Rheol. Acta,* 1975, **14**, 1077.
51. Freundlich, H. and Juliusberger, F., *Trans. Faraday Soc.,* 1934, **30**, 333.
52. Freundlich, H. and Juliusberger, F., *Trans. Faraday Soc.,* 1935, **31**, 920.
53. Freundlich, H. and Jones, A. D., *J. Phys. Chem.,* 1936, **40**, 1217.
54. Freundlich, H., *J. Phys. Chem.,* 1937, **41**, 901.
55. Freundlich, H., *J. Phys. Chem.,* 1937, **41**, 1151.
56. Freundlich, H. and Roder, H. L., *Trans. Faraday Soc.,* 1938, **34**, 308.
57. Nawab, M. A. and Mason, S. G., *J. Phys. Chem.,* 1958, **62**, 1248.
58. Mewis, J. and Metzner, A. B., *J. Fluid Mech.,* 1974, **62**, 593.
59. Maschmeyer, R. O. and Hill, C. T., *Adv. Chem. Ser.,* 1974, **135**, 95.
60. Bird, R. B., Armstrong, R. C. and Hassager, O., *Dynamics of Polymer Liquids,* Vol. I, Wiley, New York, 1977.
61. Vinogradov, G. V. and Malkin, A. Y., *Rheology of Polymers,* Mir, Moscow, 1980.
62. Vinogradov, G. V. and Malkin, A. Y., *J. Polym. Sci., A-2,* 1966, **4**, 135.
63. Han, C. D. and Yu, T. C., *Rheol. Acta,* 1971, **10**, 398.
64. Minoshima, W., White, J. L. and Spruiell, J. E., *Polym. Eng. Sci.,* 1980, **20**, 1166.
65. Oda, K., White, J. L. and Clark, E. S., *Polym. Eng. Sci.,* 1978, **18**, 25.
66. Nazem, F. and Hill, C. T., *Trans. Soc. Rheol.,* 1974, **18**, 84.
67. Montes, S. and White, J. L., *Rubber Chem. Technol.,* 1982, **55**, 1354.
68. Laun, H. M. and Munstedt, H., *Rheol. Acta,* 1978, **17**, 415.
69. Einstein, A., *Ann. Phys.,* 1906, **19**, 289; 1911, **34**, 591.
70. Jeffery, G. B., *Proc. Roy. Soc.,* 1922, **A102**, 161.
71. Batchelor, G. K., *J. Fluid Mech.,* 1970, **41**, 545.
72. Batchelor, G. K., *J. Fluid Mech.,* 1971, **46**, 813.
73. Guth, E. and Simha, R., *Kolloid Z.,* 1936, **74**, 266.
74. Batchelor, G. K. and Green, J. T., *J. Fluid Mech.,* 1972, **56**, 461.
75. Simha, R., *J. Appl. Phys.,* 1952, **23**, 1020.

76. Happel, J., *J. Appl. Phys.,* 1957, **28**, 1288.
77. Frankel, N. A. and Acrivos, A., *Chem. Eng. Sci.,* 1967, **22**, 847.
78. Krasny-Ergun, W., *Kolloid Z.,* 1936, **74**, 172.
79. Booth, F., *Proc. Roy. Soc.,* 1950, **A203**, 523.
80. Chan, F. S., Blachford, J. and Goring, D. A. I., *J. Coll. Interf. Sci.,* 1966, **22**, 378.
81. Hoffman, R. L., *J. Coll. Interf. Sci.,* 1974, **46**, 491.
82. Russel, W. B., *J. Coll. Interf. Sci.,* 1976, **55**, 590.
83. Adler, P. M., *Rheol. Acta,* 1978, **17**, 288.
84. Russel, W. B., *J. Fluid Mech.,* 1978, **85**, 209.
85. Tanaka, H. and White, J. L., *J. Non-Newt. Fluid Mech.,* 1980, **7**, 333.
86. Goddard, J. D., *J. Non-Newt. Fluid Mech.,* 1976, **1**, 1.
87. Zaremba, S., *Bull. Inst. Acad. Sci. Cracow,* 1903, 594.
88. Oldroyd, J. G., *Proc. Roy. Soc.,* 1950, **A200**, 523.
89. Oldroyd, J. G., *Proc. Roy. Soc.,* 1958, **A245**, 278.
90. Lodge, A. S., *Proc. 2nd Int. Rheol Cong.,* 1954, 229.
91. Lodge, A. S., *Trans. Faraday Soc.,* 1956, **52**, 120.
92. Rivlin, R. S. and Erickson, J. L., *J. Rat. Mech. Anal.,* 1955, **4**, 323.
93. Green, A. E. and Rivlin, R. S., *Arch. Rat. Mech. Anal.,* 1957, **1**, 1.
94. Noll, W., *Arch. Rat. Mech. Anal.,* 1958, **2**, 197.
95. Coleman, B. D. and Noll, W., *Arch. Rat. Mech. Anal.,* 1960, **6**, 355.
96. Evans, J. G., in *Theoretical Rheology,* J. F. Hutton, J. R. A. Pearson and K. Walters (Eds), Applied Science Publishers, London, 1975.
97. Hinch, E. J. and Leal, L. G., *J. Fluid Mech.,* 1975, **71**, 481; 1970, **76**, 187.
98. Bark, F. H. and Tinoco, H., *J. Fluid Mech.,* 1978, **87**, 321.
99. Schwedoff, T., *J. Phys.,* 1890, **9**, 34.
100. Bingham, E. C., *J. Wash. Acad. Sci.,* 1916, **6**, 177.
101. Bingham, E. C., *Fluidity and Plasticity,* McGraw-Hill, New York, 1922.
102. Buckingham, E., *Proc. ASTM,* 1921, **21**, 1154.
103. Herschel, W. H. and Bulkley, R., *Proc. ASTM,* 1926, **26**, 681.
104. Casson, N., in *Rheology of Disperse Systems,* C. C. Mill (Ed.), Pergamon Press, London, 1959, p. 84.
105. Hohenemser, K. and Prager, W., *Z. f. A.M.M.,* 1932, **12**, 216.
106. Oldroyd, J. G., *Proc. Camb. Phil. Soc.,* 1947, **43**, 100.
107. Slibar, A. and Pasley, P. R., *Second Order Effects in Elasticity, Plasticity and Fluid Dynamics,* M. Reiner and D. Abir (Eds), MacMillan, London, 1964.
108. Oldroyd, J. G., *Proc. Camb. Phil. Soc.,* 1949, **45**, 595.
109. Hutton, J. F., *Rheol. Acta,* 1975, **14**, 979.
110. White, J. L., *J. Non-Newt. Fluid Mech.,* 1979, **5**, 177.
111. White, J. L., *J. Non-Newt. Fluid Mech.,* 1981, **8**, 195.
112. White, J. L. and Tanaka, H., *J. Non-Newt. Fluid Mech.,* 1981, **8**, 1.
113. Suetsugu, Y. and White, J. L., *J. Non-Newt. Fluid Mech.,* 1984, **14**, 121.
114. Bergen, J. T., in *Processing of Thermoplastic Materials,* E. C. Bernhardt (Ed.), Reinhold, New York, 1959.
115. Hermann, H., *Polym. Eng. Rev.,* 1983, **2**, 227.
116. O'Connor, J. E., *Rubber Chem. Technol.,* 1977, **50**, 945.
117. Murty, K. N. and Modlen, G. F., *Polym. Eng. Sci.,* 1977, **17**, 848.

118. Lee, W. K. and George, H. H., *Polym. Eng. Sci.*, 1978, **18**, 146.
119. Crowson, R. J., Folkes, M. J. and Bright, F. F., *Polym. Eng. Sci.*, 1980, **20**, 925.
120. Chan, Y., White, J. L. and Oyanagi, Y., *Polym. Eng. Sci.*, 1978, **18**, 268.
121. Oyanagi, Y. and Yamaguchi, Y., *J. Soc. Rheol., Japan*, 1975, **3**, 64.
122. Crowson, R. J. and Folkes, M. J., *Polym. Eng. Sci.*, 1980, **20**, 934.
123. Cessna, L. C., US Patent 3 651 187, 1972.
124. Goettler, L. A. and Lambright, J., US Patent 4 056 591, 1977.
125. Goettler, L. A. and Lambright, J., US Patent 4 057 610, 1977.
126. Goettler, L. A., Leib, R. I. and Lambright, J., *Rubber Chem. Technol.*, 1979, **52**, 838.
127. Wu, S., *Polym. Eng. Sci.*, 1979, **19**, 638.
128. Menendez, H. A. and White, J. L., *Polym. Eng. Sci.*, 1984, **24**, 1051.
129. Cole, E. A., Cogswell, F. N., Huxtable, J. and Turner, S., *Polym. Eng. Sci.*, 1979, **19**, 12.
130. Ma, C. Y., White, J. L., Weissert, F. C., and Min, K., *J. Non-Newt. Fluid Mech.*, 1985, **17**, 275.
131. Ballenger, T. F. and White, J. L., *Chem. Eng. Sci.*, 1920, **25**, 1191.
132. Bagley, E. B. and Birks, A. M., *J. Appl. Phys.*, 1960, **31**, 556.
133. White, J. L. and Kondo, A., *J. Non-Newt. Fluid Mech.*, 1977, **3**, 41.
134. White, J. L. and Huang, D. C., *J. Non-Newt. Fluid Mech.*, 1981, **9**, 223.
135. Oda, K., White, J. L. and Clark, E. S., *Polym. Eng. Sci.*, 1976, **16**, 585.
136. Goettler, L. A., *25th Ann. Tech. Conf. Reinforced Plastics/Composites*, Division of Society of Plastics Industry, 1970, **14-A**, 1.
137. Goettler, L. A., *Mod. Plastics*, April 1970, 146.
138. Darlington, W. N. and McGinley, P. L., *J. Mat. Sci. Letters*, 1975, **10**, 906.
139. Owen, M. J. and Whybrew, K., *Plastics and Rubber*, 1976, **1**, 231.
140. Owen, M. J., Thomas, D. H. and Found, M. S., *Mod. Plastics*, June 1978, 61.
141. Oyanagi, Y., Yamaguchi, Y., Kitagawa, M., Terao, K. and Mochizuki, M., *Kobunshi Ronbunshu*, 1981, **38**, 285.

*Chapter 6*

# The Effects of Processing Variables on the Mechanical Properties of Reinforced Thermoplastics

LLOYD A. GOETTLER

*Monsanto Chemical Company, Akron, Ohio, USA*

## 6.1. INTRODUCTION

The addition of a second, rigid phase to a polymer in order to form a composite is attractive for either property improvement or cost reduction. In the latter case, the inexpensive particulate mineral fillers used rarely contribute to enhanced mechanical properties. Reinforcing agents require some degree of angularity. Fibers, flakes and acicular minerals reinforce along the direction(s) of their largest dimension(s). The mechanics of the stress transfer process are treated fully in Chapter 2.

The present chapter will deal with the generation and characterization of structure in a formed composite as a result of the hydrodynamics and kinematics of the fabrication process. It emphasizes the orientation, placement, and breakage of short fibers and flakes which occur during the extrusion and molding of discontinuously reinforced plastics.

The need for controlled orientation in composites, and the importance of directional orientation, are apparent from the fact that randomly placed fibers impart less than 20% of their potential modulus. Strength also rapidly declines in uniaxially oriented fiber composites as the stress angle deviates from the orientation direction.

Early work in this area has generally comprised observations of filler orientation and mechanical properties with little interpretation or analysis according to the fluid mechanics of the operation or the rheology of fiber suspensions. Few studies were of a quantitative nature and discrepancies were frequent.

This chapter will draw on the rheological behavior developed in Chapter 5 for explanation of the processing effects peculiar to these high aspect ratio composites.

### 6.1.1. Structure of Discontinuously Reinforced Composites
The incorporation of reinforcing agents produces a microscopically inhomogeneous material whose properties are more dependent upon processing techniques and conditions than are those of the polymer matrix in its unfilled form. Some elements of this structure are shown in Table 6.1 in relation to processing and properties. Sensitivity to processing increases with the aspect ratio of the reinforcement. This parameter, defined as the ratio of the longest dimension of the particle to its shortest, characterizes the efficiency of stress transfer from the matrix to the reinforcement. It becomes infinite for continuous filaments which sustain a load proportional to their relative modulus when aligned parallel to the direction of applied stress. Such a state of perfect orientation is usually not attainable in discontinuous composites. Instead, as aspect ratio increases, the properties of the composite become determined by the imperfect structure of the part, rather than by the properties of its constituent elements.

TABLE 6.1
Elements of Composite Processing–Performance Relationships

| *Process parameters* | → | *Composite structure* | → | *Composite properties* |
|---|---|---|---|---|
| Forming geometry<br>Rate<br>Temperature<br>Pressure | | Reinforcement Concentration<br>Reinforcement Size (aspect ratio)<br>Reinforcement Dispersion<br>Reinforcement Wet-out<br>Reinforcement Orientation | | Modulus<br>Strength<br>Impact resistance<br>Shrinkage |

Because of their particularly interesting behavior, attention will be focused on the relationships between processing variables, the resulting composite morphology and the ensuing mechanical properties in high aspect ratio discontinuous fiber and flake composites.

The morphological material parameters of Table 6.1 are both independent and dependent variables. Their initial state (in the feed compound) determines the nature of the material response to the processing variables. The processing, in turn, through the effects described above, causes changes to these same parameters. The most useful of these is in control of the particle orientation (i.e. mechanical anisotropy) during the fabrication operation through manipulation of the processing parameters. This provides a useful tool for optimizing performance of the composite part in regard to the stress fields imposed in use.

Emphasis will be on the large-volume application of short fibers for reinforcement. The limiting cases of continuous filaments and equiaxial particulate fillers will not be pursued because in the former case entirely different processes are used to produce much better control over composite morphology whilst, in the latter, structural effects are considerably less significant.

Although this book is directed towards thermoplastic resins, some specific examples of process–structure–property relationships will also be drawn from studies on glass fiber reinforced epoxy, polyester bulk molding compound (BMC) and cellulose fiber reinforced rubber. The same principles apply regardless of the nature of the matrix when performance is dominated by the composite structure.

For example, equivalent morphologies are encountered in thermoplastic and thermosetting systems. Mandell *et al.*[1] point out that cracks in injection moldings of fiber reinforced thermoplastics usually propagate so as to avoid regions of locally aligned fibers that are produced by the mold flow. This agglomeration of fibers has also been observed in long fiber thermoset BMC. Then the deflection of the crack according to the orientation pattern determines the composite strength. This also shows that the critical fiber length model for strength prediction is not useful when the fiber concentration is high and the fiber alignment relatively poor.

McNally et al.[2] found that better dispersion results in a stronger interface, which might lead to a stronger composite even though the fiber length is degraded. It is likely that this strengthening is related to better wet-out of the fibers and the absence of microvoids, as detected by Darlington and Smith.[3]

The optimum structure for an aligned discontinuous fiber composite would comprise long fibers that are also well dispersed in the molding compound. Unfortunately, this type of structure is very difficult to obtain. In producing a high dispersion, the fibers are broken. If the long fiber length can be preserved during the molding operation, the resulting molding will contain weak shear planes. Composite performance will then be keyed to the orientation of these planes, as described above.

The role of composite structure can be minimized by utilizing well dispersed fibers of shorter length. Tensile properties would be less directional and show a greater dependence on the characteristics of the matrix. But mechanical property development with the short glass compounds is limited by fiber length. The aspect ratio of 20–40 for these materials puts them in the range where both tensile strength and modulus are critically dependent on aspect ratio. On the other hand, an aspect ratio of 200 would yield an 80–90% modulus efficiency in most thermoplastics, depending on the fiber and matrix moduli.

In all cases, the reinforcement is coated with a coupling agent, usually a silane in the case of fiberglass (see Chapter 8), to improve stress transfer and wet-out. The data of Fig. 6.1 for a 40 vol. % 3 mm fiberglass composite with quasi-longitudinal alignment show the effect of interfacial bond strength on ultimate composite tensile strength in long fiber composites. An adhesion index has been arbitrarily defined by comparing scanning electron micrographs of transverse fractures in many samples (see Fig. 6.2). Poor adhesion is caused by high fiber bundle integrity (preventing wet-out), high resin viscosity and low molding shear or pressure. While the first two of these factors are

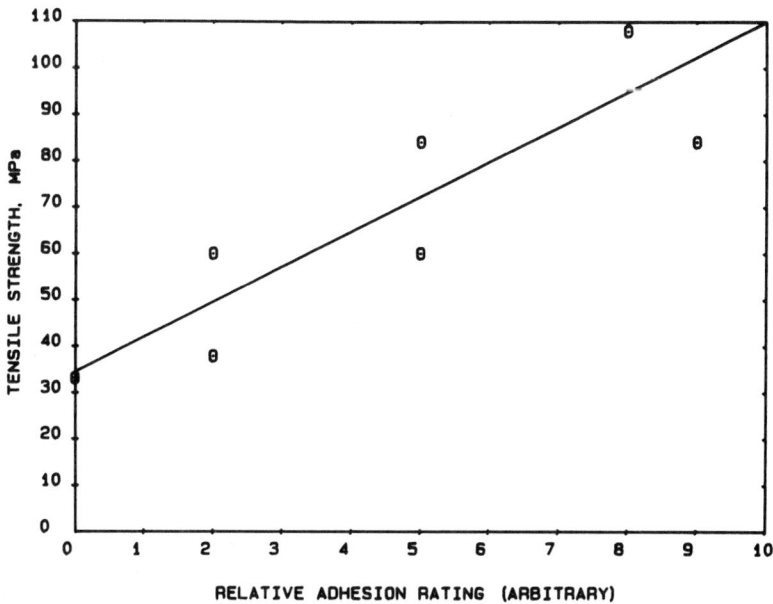

**Fig. 6.1.** Effect of fiber–matrix adhesion on tensile strength of fiberglass composites molded with a predominantly transverse fiber orientation.

material-dependent, the latter are controlled by the molding process. In this example taken from the thermoset molding area, viscosity was varied prior to molding through the degree of crosslinking in the B-staged epoxy resin. When it exceeds 50%, molecular weight is at the gel point and the resulting high resin viscosity tears apart the interface with the fibers during molding flow. Subsequent healing, after flow ceases, is impeded by the low mobility of the polymer molecules. Similar effects would be expected in a thermoplastic molding with a narrow distribution of high molecular weight molded at an insufficiently high melt temperature.

### 6.1.2. Compounding
Perhaps the first role played by processing in the determination of composite performance is in dispersion of the reinforcement during the compounding stage. Clearly, the definition of composite structure begins with the starting materials (see Fig. 6.3). Long fiberglass compounds are manufactured by coating a fiberglass roving with a

*Lloyd A. Goettler*

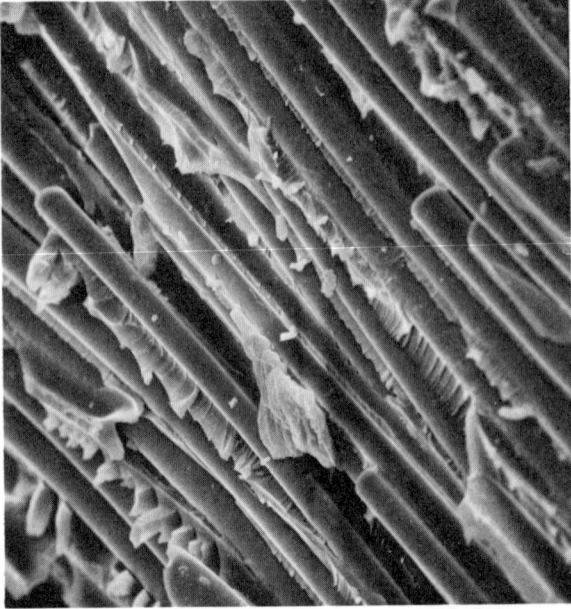

Rating: 5

Rating: 9

**Rating: 0**

**Rating: 2**

Fig. 6.2. Adhesion ratings used in Fig. 6.1. Epon® 828/methylenedianiline epoxy reinforced with 40 vol. % 3 mm fiberglass.

®Registered trademark of Shell Chemical Company.

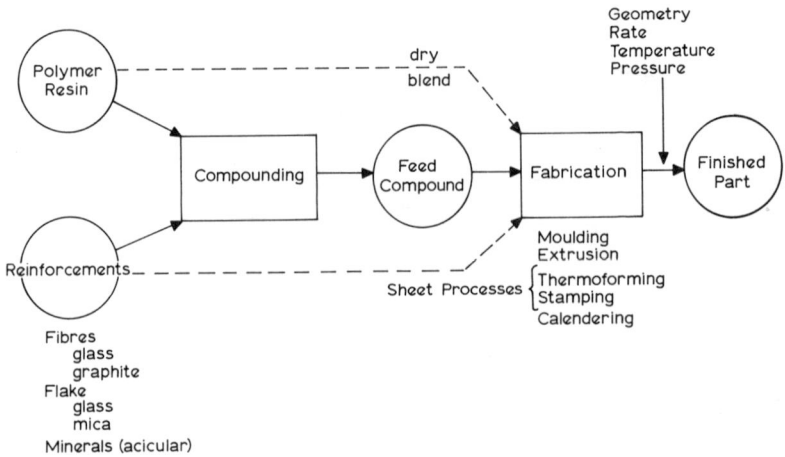

**Fig. 6.3.** Parameters affecting performance of a fabricated composite part.

thermoplastic polymer in a cross-head extruder. The product is then pelletized by chopping, usually to 10 mm length (aspect ratio ~1000). Although the glass fibers are long, equal to the pellet length, they have a poor dispersion in the polymer. The presence of a residual fiber bundle in the fabricated part would compromise strength by introducing planes of weakness. Recently, pultrusion technology has been used to produce a long fiber molding compound with improved dispersion.

A second type of reinforced plastic pellet is made by compounding the reinforcement and polymer resin together in a single or twin screw extruder.[4] In the latter, the shear and temperature history can be better controlled.[5] In multi-stage machines, chopped glass or continuous roving may be fed into an already molten polymer pool through an intermediate vent port for a gentler mixing action. However, a high degree of damage usually still accompanies the improved fiber orientation, which typically limits fiberglass length to 200–400 μm or 20–40 *l/d*. Lunt and Shortall[6] discuss the effect of extrusion compounding on fiber degradation and its effect on strength properties in short glass fiber reinforced nylon 66.

Attrition arises from the brittle nature of most high performance reinforcing materials, especially glass and graphite fiber. Exceptions are the organic fibers, either natural or synthetic. The first class comprises natural cellulose, which, as wood pulp, hemp or cotton fiber, can have a high aspect ratio,[7] above 100. Both these materials and textile fibers chopped to any convenient length (aspect ratio) offer a sufficiently high modulus and strength for reinforcing some of the softer plastics and elastomers, including thermoplastic elastomers. Moreover, their non-circular cross-section and microstructure allow considerable bending to occur without fracture during compounding and fabrication. These aspects of component material properties are covered more fully in Chapters 1 and 2.

There is thus a fundamental difference between the conditions used for the dispersion of reinforcing fibers and flakes and that of non-reinforcing particulates. Low shear, sufficient to produce wet-out without damage, is desired, rather than high shear to break up agglomerates. In the extreme, the reinforcements and the resin are not precompounded at all, but are instead fed simultaneously in free-flowing form (e.g. powdered resin) into the hopper of the final fabricating machine — injection molder, extruder, etc. Since the reinforcements then experience less processing history, their size is better preserved, and improved physical properties result.[8] Some comparative property data for different types of composites are shown in Table 6.2.

TABLE 6.2
Comparative Mechanical Property Data on Different Types of Composite Compounds

| Process | Reinforcement | Matrix | Fiber angle (deg.) | Tensile modulus (GPa) | Tensile strength (MPa) |
|---|---|---|---|---|---|
| Injection molding | 40 vol. % 3 mm fiberglass | Epoxy | 40–60 | 10·3 | 48·3 |
| Injection molding | 23 vol. % 9 mm fiberglass | Nylon 66 | 30–40 | 15·2 | 221 |
| Injection molding (controlled orientation) | 40 vol. % 3 mm fiberglass | Epoxy | 19 | 27 | 190 |
| Extrusion (prepreg) | 40 vol. % 3 mm fiberglass | Epoxy | 20 | 24·1 | 228 |
| Compression molding (aligned fibers) | 57 vol. % 3 mm fiberglass | Epoxy | 7 | 39·3 | 310 |

These data also include thermoset molding compounds to point out differences in behavior. Nylon is a superior matrix resin, providing a higher strength for the same degree of fiber stiffening. The latter correlates with fiber concentration and orientation regardless of the composite type. Careful hand alignment produces the best orientation and tensile properties, but controlled flow fabrication shows a high potential for property improvement.

Since bulk (BMC) or sheet (SMC) molding compounds contain longer (6–25 mm) fibers, both the reinforcing potential and the importance of fiber orientation can be potentially exaggerated in these materials. The higher fiber aspect ratio derives from the low viscosity of the matrix resin in its B-staged processing condition prior to full cure. Thus, although the fibers are not damaged as much by the compounding operation, they are likewise poorly dispersed, as evidenced by the comparison of composite morphologies between thermoplastic and thermoset moldings in Table 6.3. Glass beads are sometimes used to replace a portion of the short fibers in molding compounds to reduce anisotropy and viscosity,[9] and improve processability by enhancing the reinforcement packing density.[10, 11]

## 6.2. STRUCTURE–PROPERTY RELATIONSHIPS IN SHORT FIBER COMPOSITIONS

### 6.2.1. Characterization Techniques

*Fiber Length*
A complete characterization of the variable of fiber length must include determination of the distribution or frequency of occurrence of the entire spectrum of lengths in the composite. Because it is the volume fraction of reinforcement that defines the mechanical property levels of fiber reinforced composites, this distribution must be weighted according to the volume occupied by each fiber length fraction. In the case of chopped strand, wherein the diameter of all segments are the same, the weighting may be on a length rather than on a volume basis. Also, for this case, the aspect ratio distribution is identical to the fiber length distribution.

A determination of the fiber length distribution in a molded sample of a discontinuous fiber reinforced polymer can be accomplished by the following steps.

TABLE 6.3
Types of Fiber Reinforced Molding Compounds

| | Thermoplastic | | Thermosetting | |
|---|---|---|---|---|
| Molding use | Injection | Injection | Transfer | Compression |
| Manufacturing process | Compounding | Coating | Dough blending | SMC |
| Fiber length, mm | 0·1-1 | 6-9 | 3-6 | 6-25 |
| Degree of dispersion | Very good | Poor-good | Moderate | Poor |
| Wet-out | Good | Poor-good | Fair | Fair |
| Viscosity | Relatively high | Relatively high | Low | Low |
| Form | Pellets | Pellets | Wads | Wads |
| Effect of orientation on composite properties | Low | Moderate | High | Very high |
| Effect of matrix properties on composite properties | High | Moderate | Moderate | Low |

(i)     Burn the resin away from the glass fiber contained within a large piece of the sample by heating in an air atmosphere.

(ii)    Gently pull a portion of fibers, which will be interlocked to some degree, from the center of the sample, making sure that no part is within a distance from a cut edge equal to the maximum fiber length in the sample.

(iii)   If the fibers are larger than 0·8 mm in length, a dispersion may be photographed at about 6× magnification for the length measurement. For this purpose, the wad of fibers is dispersed in distilled water or a pure solvent in an ultrasonic bath to avoid further breakage. The entire sample is poured into a black (for glass) or white (for graphite fiber) enameled tray. The fibers do not agglomerate if the tray is left undisturbed as the solvent evaporates.

(iv)    In the case of smaller fibers, the photograph for length measurement must be taken at 20–50× magnification under a microscope. The fibers are dispersed in an ultrasonic bath, as above, but are then poured on to a microscope slide. The sides can be built up to avoid spillage. After the solvent evaporates, the fibers are photographed under dark field illumination. This yields a photograph of the dispersion with excellent contrast. A dispersion of graphite fibers could instead be collected on a piece of white filter paper and photographed under incident illumination.

In either of steps (iii) or (iv) it is necessary to make a quantitative transfer of the suspension of fibers in the dispersing solvent to the evaporating surface(s) in order to ensure a homogeneous sample. Each suspension may be divided on to more than one evaporation surface if it is ensured that all parts of all dispersions are photographed uniformly.

(v)     A border is drawn around the edge of each photograph to leave a margin equal in width to the longest fiber length. All fibers of which part lies within the interior of the frame must be measured for length. This measurement must be weighted according to the length of the fiber falling within the interior region. (This procedure assumes that all fibers have the same diameter. In the case of whiskers, for example, the width would have to be measured as well, and weighting would be according to the particle volume that falls inside the framework.)

(vi)    The length measurements may be made from the photographs using a particle size analyzer of any manufacture. Computerized analyzers are now available

(vii)   By using a large field and selecting the border through the separations around large groups of fibers so that no fibers cross the border, the weighting length will equal the actual fiber length in all cases. An example is shown in Fig. 6.4.

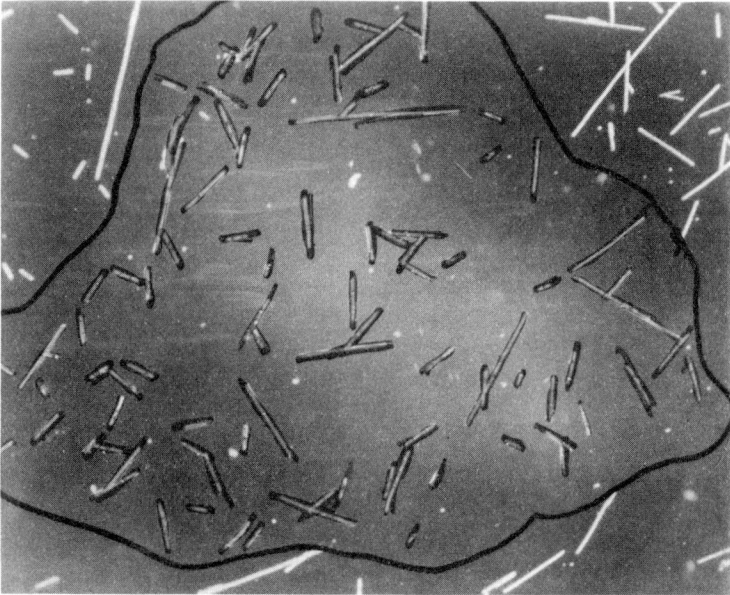

**Fig. 6.4.** A dispersed field of short glass fibers removed from an injection molding of 40 wt % (23 vol. %) reinforced polyethylene terephthalate, prepared for length characterization. The curved line defines the field of measurement in this photograph.

(viii)  At least 500, and preferably 1000, fibers should be measured for each material characterized. The length-weighted length distribution can easily be calculated by computer.

Some examples of measured fiber length distributions are shown in Fig. 6.5; $F_v$ is the volume-weighted probability density function. For example, 70% of the volume of fibers in the sample have a length that is less than that corresponding to $F_v = 0.7$. The fiber lengths used in these

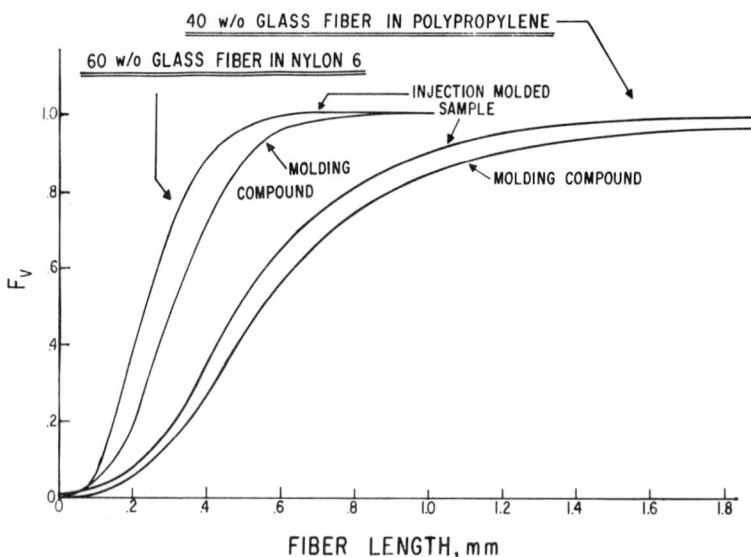

**Fig. 6.5.** Measured fiber length distributions in two thermoplastic molding compounds and their associated injection moldings.

melt blended thermoplastic molding compounds are rather small, especially at the higher volume content of the nylon resin. However, little further damage results on passage through the screw of an injection molding machine. Table 6.4 gives a comparison of the weighted median fiber lengths in different molding compounds. Note the considerably greater length, both initial and retained, in the thermosetting epoxy formulation. This derives from the gentler compounding action in conjunction with small shearing stresses developed during transfer molding in comparison with the faster injection molding of the more viscous thermoplastics.

### Orientation Distribution

The most economical means for the large-scale fabrication of hardware from short-fiber composite materials are flow processes, such as injection molding or extrusion. Fiber orientation resulting from the flow is one of the major factors determining the mechanical strength as well as the stiffness of the molded part. When a composite is fabricated

TABLE 6.4
Fiber Lengths in Thermoplastic and Thermoset Moldings and Molding
Compounds

| | *Fiber content* | *Median fiber length (mm)* | |
|---|---|---|---|
| *Material* | *(wt %)* | *Molding compound* | *After molding* |
| Nylon 66 | 60 | 0·23 | 0·23 |
| Nylon 6 | 60 | 0·30 | 0·23 |
| Polypropylene | 40 | 0·56 | 0·49 |
| PET | 40 | 0·36 | — |
| PET | 20 | 0·43 | — |
| Epoxy | | | |
| With Binder | 60 | 3·1 | 3·0 |
| Heat-conditioned | 60 | 3·1 | 1·6 |

by molding or extrusion, the structure is neither well aligned nor random. A tendency toward good alignment can be obtained by tailoring the flow geometry of the part, but significant angular deviations always exist among the fibers, so that the uniaxial elasticity equations cannot be accurately applied.

This importance of orientation as a morphological parameter describing composite structure and performance leads immediately to the need to measure it. For example, variations in the orientation distribution could be monitored as a quality control check on subsequent strength and stiffness. In addition the development of process control schemes for fiber orientation and placement must rely on adequate means for *completely* characterizing the distribution.

A complete characterization of orientation is emphasized, rather than a mere representation of degree of alignment in some mean direction, because orientation of itself is inconsequential; it is useful only as a predictor or indicator for mechanical properties. The calculation of these properties requires a summing of contributions across the entire spectrum of orientations present in the composite part. Non-linear functional relationships between the mechanical properties and the fiber directionality preclude a meaningful relationship of *average* values. The single-valued orientation functions defined by White and Spruiell[12] are convenient descriptors of the character of the orientation distribution. Pipes *et al.*[13] show how multiple functions can better represent the distribution in calculating mechanical properties.

In any molding containing short fibers, orientation distributions of two types may be encountered. These are a distribution in the angles assumed by individual fibers at any one macroscopic location in the molding, and the overall variation with position in the molding. The former is related to the 'spread' in the distribution, and would be unlikely to approach zero in any short-fiber composite material. On the other hand, the overall variation may be considered in terms of changes in the local average direction. These concepts are useful in classifying the types of orientation measurements which are used and they also help to conceptualize complex orientation patterns. Thus, these may be specified as distribution functions in which the parameters, such as the mean or the standard deviation, are functions of coordinate position in the molding.

One of the major obstacles to measuring a complete orientation distribution in a piece is the need for a three-dimensional characterization. Planar angles must be limited to sheets (or thin members) whose thickness is less than a small fraction of the fiber length. Otherwise, the actual fiber position relative to a fixed direction, such as that in which the stress will act or that normal to a biaxial stress field, could differ significantly from that measured by the projection of the fiber in a plane. A second complication is the need, for the above-mentioned reasons, to avoid averaged measurements alone, and also those measurements, average or individual, of physical properties which are not simply (linearly) related to orientation. Although these types of measurements may impart some indication of the orientation direction, they cannot be employed to calculate other physical-mechanical properties. Methods which measure angles directly, and preferably in three dimensions about an axis, are consequently the most useful.

*Methods involving polished sections.* These involve microscopic examination of planes cut into the material in which the individual fibers, or groups of fibers, are made visible by some mechanical or chemical treatment, such as etching or staining. The orientation in a direction normal to this plane can be determined either by examining planes in a different direction, or from the shapes of the fiber cross-sections in a single plane.

(a) *Normal sections.* We consider first a method utilizing normal planes cut into the composite specimen.[14] This analysis is most accurate and easiest to perform when most of the fibers are lying in a position more nearly in the direction of interest than opposed to it. It is a simple but meaningful method of measuring and combining *planar* angles that can predict the distribution of *polar* angles, $\theta_1$, that the fibers make with a stress axis and the corresponding mechanical stiffness. The angles designated $\phi_2$ and $\phi_3$ are measured in two orthogonal planes that include the axis and these are then combined to calculate the local and overall average polar angles for the molding or the stiffness coefficients. If the molding has axial symmetry, and the gate location is also symmetrical about the axis, the resulting orientation pattern will be symmetrical, although macroscopically non-uniform. If the end gate is small, the structure of the molding usually comprises a core of transversely oriented fibers surrounded by envelopes of orientation either random or parallel to the flow direction. A brief description of the procedure follows.

In order to describe orientation as a function of position, two coordinate systems must be defined.

(i) Any coordinate system can be used to specify position in the part; rectangular coordinate directions for a typical bar molding are shown in Fig. 6.6. (The orientation patterns also shown will be explained in Section 6.3.2.) The origin is located in the center of the bar at the gate; $x_1$ is the flow direction during molding or the stress direction during tensile testing and increases in the flow direction. Since there are two transverse directions in a bar, $x_2$ is arbitrarily directed upwards across the narrow dimension of the bar; $x_3$ is across the wide dimension in a direction consistent with a proper right-hand coordinate system. These rectangular coordinates should be normalized so that

$$0 \leqslant x_2, x_3 \leqslant 1$$

(ii) The orientation at a point can be specified by two angles of one or more spherical coordinate systems. Each system, using one of the three rectangular coordinate directions as its polar axis, comprises a latitude angle, $\theta_i$, about the axis, and a longitude angle, $\phi_i$, specifying the angle projected in the plane normal to that polar axis. Since the longitude angles, $\phi$, are defined in a plane, they are easier to measure from planar surfaces cut into the sample than is $\theta$. However, it is the latter quantity that relates more directly to the anisotropic physical or mechanical properties

A: Bar

$\phi_3$ (Planar)
or
$\theta_1$ (Polar)

B: Rod

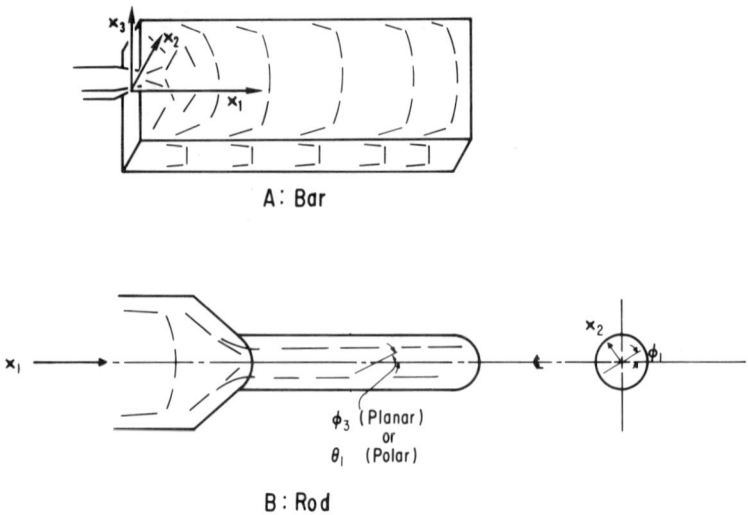

**Fig. 6.6.** Coordinate geometry and resulting fiber orientation patterns in two simple tool geometries representing diverging and converging flow fields.

along an axis in the material. A relationship between the angles allows the calculation of any angle from the measurements of two others. For the generalized system illustrated in Fig. 6.7, the angles are related by

$$\tan^2 \theta_1 = \cot^2 \phi_2 + \cot^2 \phi_3 \tag{6.1}$$

where $\theta_1$ is identified with the angle to be used in the elasticity transformation equations. An orientation completely in the flow direction is described by $\phi_2 = \phi_3 = 90°$ or $\phi_1 = 0°$. In any three-dimensional part in which the smallest dimension exceeds a fraction of the fiber length, a two-dimensional analysis using $\phi_2$ or $\phi_3$ alone as the orientation angle could be seriously in error.

There are two parts to the overall orientation distribution.

(i) The orientation over a small neighborhood, usually corresponding to an opened bundle of fibers, as a smoothed function of position in the cross-section. This part is designated $\overline{\phi}_2(x_2, x_3)$ or $\overline{\phi}_3(x_2, x_3)$ where the single bar is used to indicate a local mean value.

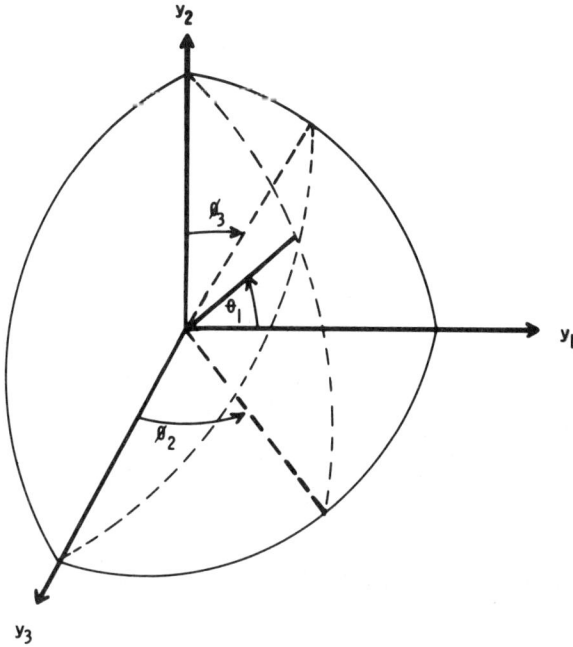

**Fig. 6.7.** Coordinate system for fiber orientation relating angles $\phi_2$ and $\phi_3$ measured in planes cut parallel to the major direction $y_1$ to the polar angle $\theta_1$ around $y_1$.

(ii) The scatter of individual fibers about this local mean value. Since this quantity is relatively constant for each particular type of molding, an average value can be used. The distributions of deviations of individual fiber orientations from the local mean can be measured in high magnification photomicrographs of polished surfaces cut parallel to the $x_2$ and $x_3$ directions. Typical standard deviations for individual fiber scatter in these planes range from 10 to 15°.

The measurement of $\bar{\phi}_2$ and $\bar{\phi}_3$ can be directly made from low-magnification photographs of planes normal to the $x_2$ and $x_3$ directions cut along a length of the sample. These can be either optical or X-ray radiographs. For the first case, the surfaces must be ground smooth, etched with an agent that mildly attacks one of the phases, and, if necessary, stained. This technique is particularly effective with long fibers, which tend to lie in bundles. With radiography, thin sections must be cut from the part if the variation in orientation in the normal direction is to be detected, and tracer fibers should be used. By either method, the local orientation appears as a single mark that indicates the mean fiber direction. An example of a surface obtained for the measurement of $\bar{\phi}_2$ is shown in Fig. 6.8. The angles are quickly measured with the protractor scale on a drafting instrument.

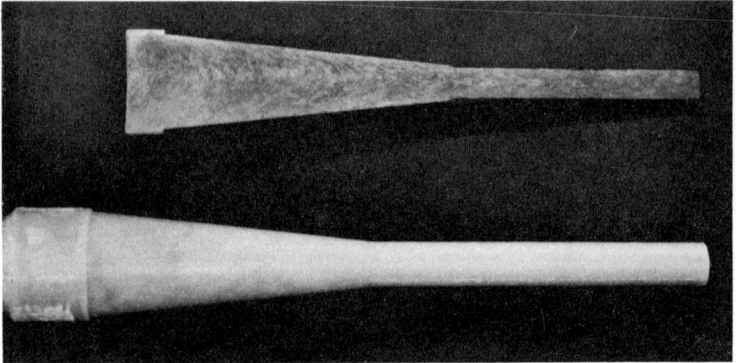

**Fig. 6.8.** A rod molding with a prepared section showing the generation of longitudinal fiber orientation in a converging flow.

In order to characterize the orientation or predict property levels for the entire molding it is necessary to average the resulting local values over the cross-sectional area. A smooth representation of the mean planar angles $\bar{\phi}_2$ and $\bar{\phi}_3$ as a function of position can be obtained by fitting the data to a polynomial expression in the coordinate distances using a multilinear regression technique. The standard error of the regression is taken to be the standard deviation of random variation in $\bar{\phi}_2$ or $\bar{\phi}_3$ about the smoothed distribution over the distance coordinates. These values must be combined with the individual fiber variances to yield the overall standard deviation of the orientation distribution.

Thus, the distributions of the planar angles themselves can be further resolved into local mean directions $\bar{\phi}_2$ and $\bar{\phi}_3$, which are smoothed functions of coordinate position in the part, and the local distributions of individual fibers around these means. That is, the distributions of the fibers are taken to be normal with mean $\bar{\phi}_2(x_1, x_2, x_3)$ or $\bar{\phi}_3(x_1, x_2, x_3)$ and standard deviations $s_2(x_1, x_2, x_3)$ or $s_3(x_1, x_2, x_3)$.

The average axial angle $\bar{\theta}_1(x_2, x_3)$ cannot be calculated directly from $\bar{\phi}_2$ and $\bar{\phi}_3$ according to eqn (6.1), which applies only to individual fiber orientations and not to the means of the distributions, but a numerical integration can be made over their distributions. The small variations of the individual fiber orientations around the local means can be taken to a first approximation to be independent. With this simplification, the statistical formula for the mean of a function of two random variables can be applied to $\bar{\theta}_1(\bar{\phi}_2, \bar{\phi}_3)$ at each point $(x_2, x_3)$:

$$\bar{\theta}_1(x_2, x_3) = \int_{-\infty}^{\infty} \int_{-\infty}^{\infty} \theta_1(\phi_2, \phi_3) N(\phi_2; \bar{\phi}_2, s_2) N(\phi_3; \bar{\phi}_3, s_3) \, d\phi_2 \, d\phi_3$$

(6.2)

where the $N$ are normal distributions and $\theta_1(\phi_2, \phi_3)$ is given by eqn (6.1). Similarly, the local average stiffness or compliance in the direction of stress, designated $\overline{W}_{11}$, depends only on $\theta_1$ and is given by

$$\overline{W}_{11}(x_2, x_3) = \int_{-\infty}^{\infty} \int_{-\infty}^{\infty} W_{11}[\theta_1(\phi_2, \phi_3)] N(\phi_2) N(\phi_3) \, d\phi_2 \, d\phi_3 \qquad (6.3)$$

In this equation $W_{11}[\theta_1]$ designates the angular dependence of the stiffness element for a well-aligned composite.

To obtain overall average angle or elasticity coefficient, the local parameter must now be averaged over the cross-section, for example

$$\overline{\overline{W}}_{11} = \int_A \int W_{11} \, dA \qquad (6.4)$$

In addition, the Young's modulus can be given by

$$\overline{\overline{E}}_{11} = \overline{\overline{C}}_{11}/P \qquad (6.5)$$

where $\overline{\overline{C}}_{11}$ is the longitudinal stiffness and the Poisson correction term is given by

$$P = (1 - v_{23})/(1 - v_{23} - 2v_{12}v_{21})$$

Since $\theta_1$ only takes on positive values in the range $0° \leqslant \theta_1 \leqslant 90°$, its average, $\overline{\theta}_1$, is a measure of the degree of dispersity in symmetric orientation distributions and can equal zero only in composites that are perfectly aligned in the stress direction.

This method is believed to simplify conceptually the study of orientation by breaking the types of variation down into microscopic and macroscopic categories. In addition, it inherently considers the positions

*Lloyd A. Goettler*

of a great many fibers in determining the orientation direction, but may require only a relatively small amount of measurement on individual fibers to determine the individual fiber scatter variance. For these reasons, it is preferred to the alternative procedures of tediously analyzing the polar angles of many individual fibers or of only a few tracer fibers. The method is practical, if not exact.

Some results obtained by the procedure are shown in Figs 6.9 and 6.10. The tensile strength and modulus of pieces molded from a composite reinforced with 3 mm fiberglass are correlated with the average angle determined by the above technique. A one-to-one correspondence is noted for these bars and rods with end gates that generate substantially different orientation in the core of the molding.

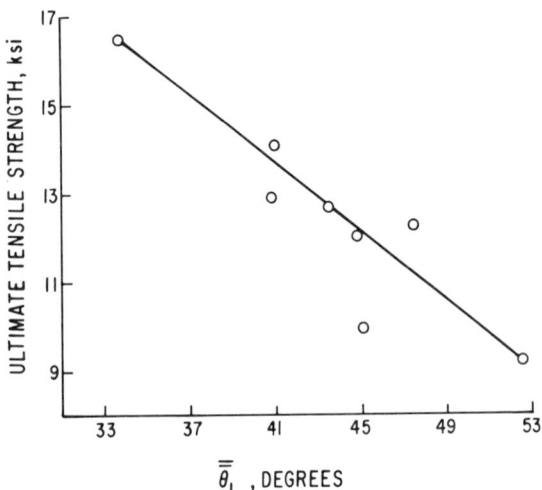

Fig. 6.9. Tensile strength of a composite molding related to the average fiber orientation angle as characterized by the technique of planar measurements in perpendicular polished sections.

This method is direct and yields quantitative estimates of the important distribution statistics. However, it is destructive, and although the orientation measurements are time-consuming, the automatic measuring and counting of orientation distributions directly from photomicrographs of sample surfaces by image-analyzing

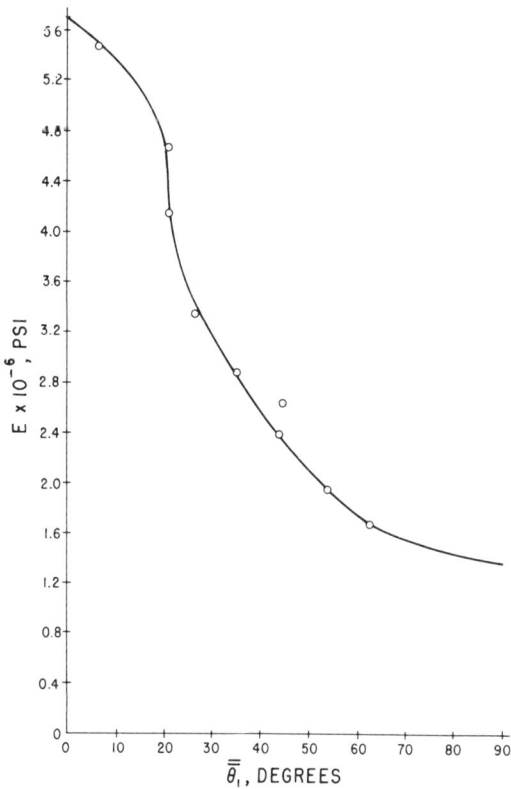

**Fig. 6.10.** Measured Young's modulus related to the same average fiber orientation angle as in Fig. 6.9.

computers should be feasible. It is versatile and can calculate those mechanical properties whose relationship to orientation in a highly aligned composite of the same materials is known or calculable. The method is, unfortunately, restricted to 'long' discontinuous fibers ($l > 1$ mm), for only these are sufficiently bundled to show gross orientation patterns under low magnification.

(b) *Single sections.* In this method of analysis, a single plane is cut through the sample and highly polished for photography at a high magnification (generally $>200\times$ for fiber diameters $<15\,\mu$m). The cross-section of circular fibers will appear as ellipses whose eccentricity increases as the angle they make with the normal to the plane increases.

This method is exact only for elongated reinforcements whose cross-section is circular or axisymmetric with a high symmetry. For irregular cross-sections, such as are observed in carbon fibers, the degree of uncertainty in angle decreases as the size of the surface irregularities with respect to the effective diameter decreases. A typical field is illustrated in Fig. 6.11.

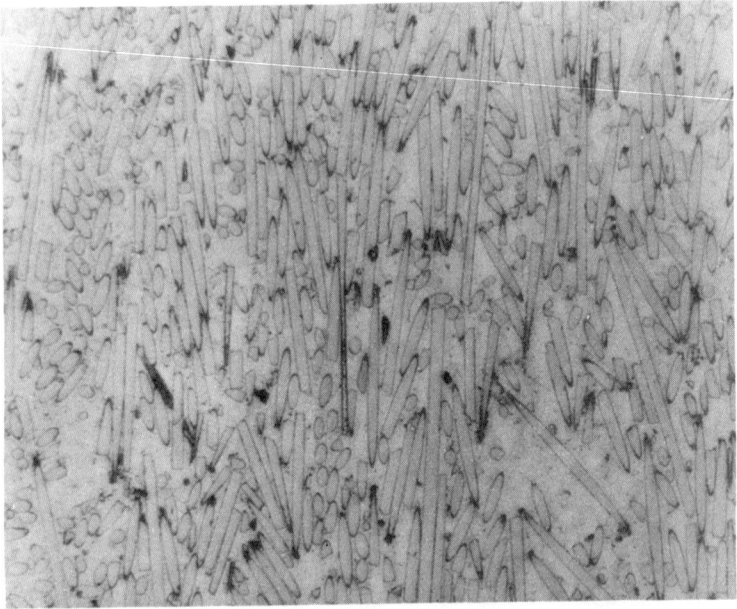

**Fig. 6.11.** Fiber cross-sections in a polished plane with ellipticity showing angle of inclination.

In particular, the polar angle about the surface normal, $\theta_1$, is related to the major and minor axes of the fiber section $a$ and $b$ according to

$$\cos \theta_1 = b/a \qquad (6.6)$$

This relationship is useful only in the range $30° < \theta_1 < 60°$, wherein the cosine function changes rapidly with its argument. Thus, the method is restricted to those situations in which the average angle of the reinforcement in a composite happens to be within the appropriate range to the stress axis of interest. A further limitation is the tedious nature of the measurements, which must be made on hundreds of individual filaments. Multiple locations would normally have to be investigated to determine macroscopic orientation variations as well. Of course, the labor is drastically reduced again here by the use of an image-analyzing computer. The programming of such an instrument for this application is more direct than for the methods discussed above under 'Normal sections' because individual fibers are clearly distinct at high magnification. The major advantage of this technique over those previously described is that it yields the polar distribution of individual fiber orientation directly and this is the most useful form for 'orientation information'. To account for macroscopic variation, the distribution parameters could, as before, be taken as functions of coordinate position.

*Microwave.* When microwaves are passed through a composite material, the presence of the second phase causes changes in amplitude and phase of the transmittal signal which can be related to the reinforcement content, the density, the presence of flaws, etc. However, in addition, when the source is made to emit linearly polarized waveforms, the signal strength becomes a function of the angle between the fiber orientation and the sensor. Rotation of the sensor relative to the sample will then yield the mean orientation direction. It is apparently also possible to get some measure of the orientation distribution, or 'degree' of orientation. The microwave techniques have been applied to both conducting and insulating fibers embedded in a nonconducting matrix. For further details the reader is referred to the literature.[15, 16] The equipment required for these techniques tends to be expensive but not so much as for an image-analyzing computer.

*Radiography.* This qualitative technique developed by Darlington[17] involves the incorporation of some fibrillar tracer reinforcement that is opaque to the electromagnetic radiation, for example lead silicate fiber. Since the pictorial radiographs display only a two-dimensional representation of the orientation, the method would be limited to flat sheets unless normal views were taken at each location and combined according to the methods set forth in the section on 'Methods involving polished sections'. Measurement would be facilitated by the good contrast of the few tracer fibers, but some uncertainty might be introduced in using them to indicate the orientation of the primary reinforcement. Because of the low tracer fiber density, precise point estimates of the orientation distribution would not be possible. Thus, tracer analysis is not suited to complete quantitative orientation analysis when both microscopic and macroscopic variations occur. However, at least for thin specimens when the tracer fibers could not interfere with the material performance, radiography would serve as a rapid qualitative NDT technique. Applications are described by Crowson *et al.*[18]

*Scanning electron microscopy.* Scans at low magnification ($\leqslant 50\times$) may be used to indicate qualitatively the overall orientation pattern. At higher magnification ($500\times$), the dispersion in individual fiber angles is evident. These views should be made on the surface of a fracture, but if the matrix is too ductile, clean breaks will not occur and the fiber orientation may also be altered in the drawing process. A better indication of fiber arrangement occurs when the bonding between the fiber and the matrix is low, since longer sections of the fibers then extend out from the surface. Alternately, the matrix can sometimes be removed by burning. The scanning electron microscope should be used for a very approximate determination.

*X-ray diffraction.* The technique of X-ray diffraction for determining the polar orientation of fibrous reinforcements in a matrix is a special method that is limited to crystalline reinforcements. As an example of the application of this technique, Schierding[19] has measured the pole densities of X-ray reflections from $\alpha$-SiC whiskers embedded in various polymer matrices. By assuming the pole density distribution to be a normal bivariate distribution, the standard deviations of the principal normal curves in two orthogonal directions, $\sigma_x$ and $\sigma_y$, can be calculated from measured X-ray intensity distributions. These can be related to the polar orientation distribution. The method has recently been extended to encompass crystalline polymer fiber reinforcements by Menendez and White.[20] The equality between X-ray intensity and orientation distribution occurs because the former is directly proportional to the volume fraction of reinforcements exhibiting the proper angular orientation. Accuracy is $\pm 10\%$.

*Light diffraction.* McGee and McCullough[21] have perfected a Fraunhofer diffraction technique for measuring the state of two-dimensional short fiber orientation from radiographed masks of the composite structure.

*Sonic modulus.* Directional measurements of the elastic modulus by monitoring the propagation velocity of sound waves through the composite can provide some information on the fiber orientation pattern.

## 6.2.2. Mechanical Relationships for Elasticity and Failure

An analysis for the tensile behavior of short fiber composites usually begins with the strength and modulus equations for an aligned array. These equations point out the dependence upon reinforcement concentration and aspect ratio alone (in addition to the constituent properties). However, implicit in these equations is the assumption that the fibers are well aligned in the stress direction (a condition that is extremely difficult to attain in short fiber composites), and that failure occurs either by fiber fracture or by fiber pullout. However, the quantity $\tau$ relating to the strength of the interface is determined by both the degree of wetting and adhesion between the fibers and the matrix resin. These in turn are subject to the conditions under which the composite was fabricated and may be affected either beneficially or adversely. If the dispersion of long fibers is incomplete, cracks may easily propagate parallel to the fiber orientation regardless of the stress direction.

Finally, composite failure may occur through molding flaws, or the presence of excessive voids. Thus, the full picture is considerably more complicated than the idealized equations of composite mechanics indicate.

In general, composites containing long glass fibers are especially subject to problems of dispersion, overlapping of fiber ends, and orientation, whereas the short-glass compounds are limited by fiber length.

In some special situations it may be possible to estimate closely both the elastic and ultimate properties of a practical composite. An example of this is in injection moldings made from a brittle polymer reinforced with a high ($>25$ vol. %) concentration of ($>1$ mm) glass fibers. In such composites a substantial but imperfect realignment of the fibers is produced by the flow during the molding process. Since the fibers are not completely dispersed and overlapped in these long-fibered molding compounds, fracture preferentially occurs by crack propagation along the direction of fiber orientation. The localized regions of differing orientation provide a continuous fracture path across the sample. The fracture surface is then a curvilinear plane that parallels the local fiber orientation. Few fibers are actually broken and most lie within or at a small angle to the surface.

The ultimate tensile strength of off-axis uniaxial composites exhibiting this fracture mechanism has been shown by Ishai and Lavengood[22] to depend only on the transverse tensile strength and fiber angle according to the semi-empirical equation:

$$\sigma_u = \frac{\sigma_{u, trans}}{\sin \theta} \tag{6.7}$$

This equation can be applied to a practical composite containing a distribution of fiber angles by summing the contributions of each angular element to the overall strength. The factor $1/\sin \theta$ must be averaged by integrating over the measured distribution of fiber angles. For some typical orientation patterns in 12 mm diameter end-gated molded rods, tensile strengths of ~125–185 MPa are predicted from a transverse strength of 40 MPa. Experimentally measured strengths of these samples fall in the range 125–165 MPa. By taking the strength of the region of worst orientation in these composite samples, where the crack would initiate, instead of the average orientation factor as outlined above, a poorer lower bound estimate of 115–160 MPa obtains.

By integrating over the measured orientation distribution using methods given (for example) by Goettler and Lavengood,[23] it is possible to predict Young's modulus of a molded composite, knowing only the elastic constants for the specially orthotropic well-aligned model specimen of the same material and fiber content. Some data showing the effectiveness of these procedures are given in Fig. 6.12. These are for rectangular bars gated at one end. Most of the discrepancy can be attributed to the neglect of certain Poisson effects in the analysis. An aspect ratio of 200 gives the same results as if the fibers were infinitely long (continuous).

A typical orientation distribution for one of these moldings is shown in Fig. 6.13. Note that it is complex, showing three modal values. These perhaps correspond to the transverse core, the longitudinal skin and a highly sheared region between them.

## 6.3. SUSPENSION RHEOLOGY RELATED TO PROCESSING

An understanding of the rheology of dispersed systems is required for interpretation of the flow phenomena occurring during the fabrication of reinforced plastics. Phenomena such as die swell, pressure drop, flow

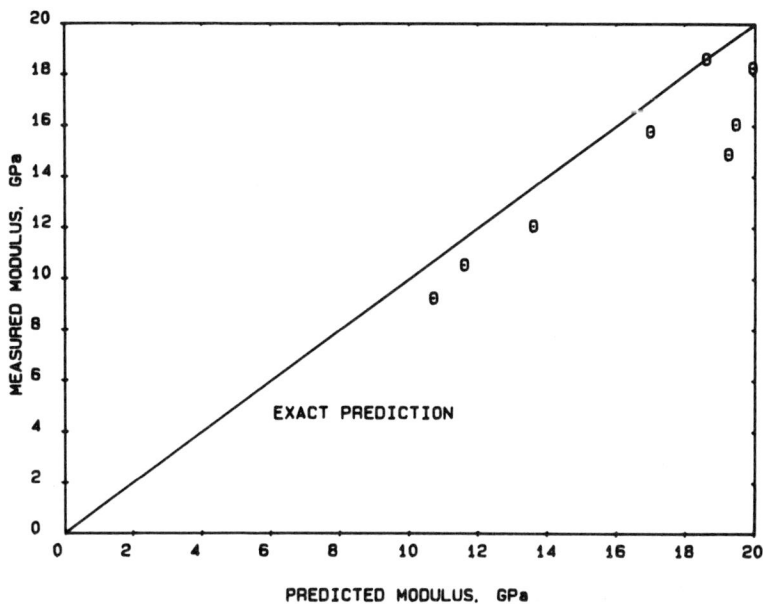

**Fig. 6.12.** Measured Young's modulus in composite moldings compared with predictions made from eqn (6.5).

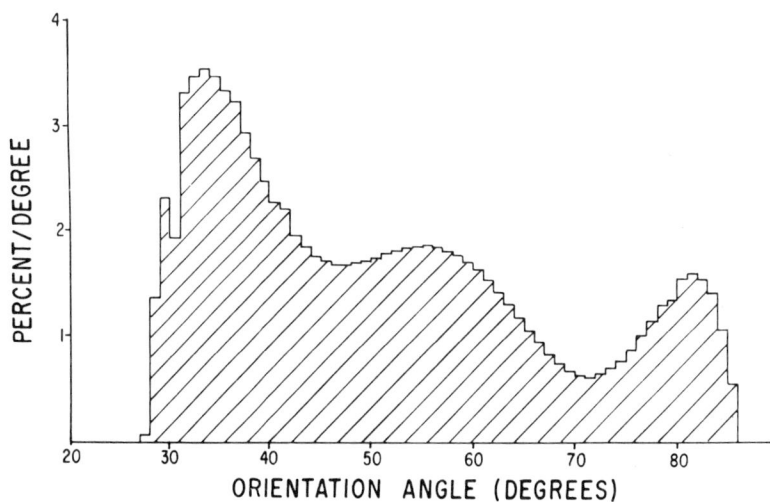

**Fig. 6.13.** Measured distribution of the orientation angle $\bar{\theta}_1$ in molded composite bars of 3 mm fiber.

rate and flow instabilities depend upon the viscosity and elasticity of the suspensions. However, since molding systems are complicated mixtures of irregularly shaped particles suspended at high concentration in a viscoelastic matrix, they have received little experimental attention.

While some advances have recently been made in a complete mathematical formulation of the overall problem,[24, 25] it is not surprising that the initial approaches have been semi-empirical and pragmatic.

It is well known that the addition of particles to a suspending medium will cause an increase in viscosity. Following the simple theoretical equation derived by Einstein for a low concentration of spherical particles in a Newtonian liquid, various empirical and theoretical forms have been proposed for the shear viscosity of suspensions to extend the range of applicability. These are described in Chapter 5.

Some of these equations may also be applied to the more complex cases of particulate fillers in polymer melts. The addition of particulate matter to a polymer melt increases the viscosity in a regular way at all shear rates. Furthermore, the addition of filler reduces elasticity, and hence post-extrusion die swell decreases dramatically.

Viscosity also increases as the concentration of reinforcing fibers increases. In elongational flows, viscosities for fiber suspensions are greatly in excess of the factor of three times the shear viscosity expected for Newtonian fluids at low elongational rates. Consequently, the pressure drop in entrance flows of fiber suspensions are correspondingly high. However, the elongational viscosity drops rapidly at increasing rates of deformation, such as are predominant in molding and extrusion flows. Similarly, the shear viscosity increases more rapidly with fiber concentration at low shear rates, and only slightly at the high shear rates that would be generated in injection molding.

Fiber suspensions in thermoplastic polymers follow the power law model with a low plasticity index (in the range $\frac{1}{3}$–$\frac{1}{2}$), which suggests an approach to plug flow (highly flattened velocity profile). Shear flow becomes limited to a thin ($<0.5$ mm) layer richer in resin near the wall. There is then no effect of shear on the orientation of the reinforcing fibers. The tendency to plug flow would be expected to be smaller in thermoplastic systems because of the shorter fiber length of the reinforcement and the higher viscosity of the matrix in comparison with thermosets.

Whereas fiber suspensions tend toward plug flow in shear, they are highly responsive to elongational flows. The tendency for fibers to align along the direction of stretching is a well-known phenomenon. Conversely, when the suspension is compacted, the fibers turn normal to the streamlines. The degree of response is determined primarily by the length of the fibers and the amount of deformation. The importance of elongational flows in polymer processing is currently being recognized. Some examples of such flows that occur in extrusion and injection molding are simple shear and extension. In these elongational flows, the entire medium is deformed rather than just the outer layer of the fiber suspension in shear.

## 6.3.1. Orientation Behavior of Spheroidal Suspensions

The most significant results of rheological studies regarding their applicability to composite fabrication deal with the orientation behavior exhibited by the fibers in suspension.

Extensive studies of the motion of single fibers in dilute suspensions have been carried out by Mason and coworkers.[26] They found for shear flow in a pipe that prolate spheroids (and rods) take on a tumbling motion in an orbit defined by the initial position of the fiber in the flow. The rate of rotation is greatest when the fibers are normal to the flow, so that on a time average the particles tend to align with the flow direction. In slow shear flows there is no migration of rigid particles in Newtonian liquids, but in viscoelastic materials they migrate from the wall to the region of flat velocity profile at the center.

The concentration of the reinforcement in a polymer composite is usually sufficiently high for interparticle interactions to become predominant determinants of rheological behavior. Folgar and Tucker[24] have shown that the critical volume concentration for short fiber interactions is very low, and decreases further as the aspect ratio of the fiber increases, e.g. $0.3$ vol. % at $16 \, l/d$ and $0.015$ vol. % at $84 \, l/d$. Moreover, interactions may also occur with the walls of the flow channel or mold. These may be mechanical, since fibers at an angle to the wall are excluded from a boundary region equal to half their normally projected length, or hydrodynamic, in the sense of an anomalous shear layer in the vicinity of the wall.

Both types of interactions manifest their strongest effects upon the local orientation vector and upon aspect ratio degradation. Fibers and platelets tend to lie parallel to a solid boundary in order to reduce the size of their excluded volume, whereas interparticle interactions greatly modify the rheological response of the suspension.

At higher concentrations, where plug flow occurs with shear being limited to a thin lubricating layer near the wall, the rods in the central (unsheared) core do not rotate. This phenomenon of plug flow has also been observed by Bell[27] in simulated epoxy molding resins reinforced at high loading (50 vol. %) with glass fiber. It has an important consequence to the orientation of long fibers in injection moldings.

Takano[28] has extended these studies to include fiber orientation and other flow effects in highly concentrated suspensions. In addition to fiber concentration, the effects of die geometry and flow rate on the migration and rotation of short fibers in a $5 \cdot 5 \, \mathrm{Nsm}^{-2}$ suspending medium were studied cinematographically by causing model systems incorporating a low concentration of tracer fibers to flow through narrow uniform and convergent rectangular channels. These tracers could be viewed optically because the index of refraction of the great majority of the fibers was matched to that of the liquid.

His most important conclusions are as follows.

(i) In concentrated suspensions ($\geqslant 5$ vol. %) of short fibers in a low viscosity resin ($5 \cdot 5 \, \mathrm{Nsm}^{-2}$), a plug flow is observed for fibers in both uniform and convergent rectangular channels.

(ii) The longitudinal alignment of fibers occurs exclusively in convergent channels. The degree of fiber alignment is decreased when the translational motion of fibers is restricted by the fiber–wall and fiber–fiber interactions.

(iii) Two kinds of flow instabilities are observed; one is the blockage of narrow channel exits with fibers due to the fiber–wall interaction and the other is the fluctuation in the streamline due to the fiber–fiber interaction during convergent flow.

Harris and Pittman[29] also studied fiber rotations — in very dilute solutions flowing through convergent channels. Approximately the same results and degree of fiber orientation were obtained as had been seen by Takano using higher fiber concentrations. Additional visualization measurements of fiber rotations in concentrated suspensions have been pursued by Lee and George[30] following the work of Takano. These and other different investigations are covered by McNally[31] in a general review of short fiber reinforced polymers.

The resin viscosity is found to have a significant effect on the flow behavior of these concentrated suspensions of 3 mm fibers. When the resin viscosity exceeds about $100 \, \mathrm{Nsm}^{-2}$ the velocity profile observed in a uniform channel is not flat, but resembles that for a pseudoplastic fluid, being rounded, but short in center. Then the longitudinal alignment of fibers can occur in uniform channels as well, by a shear mechanism.

In practice, thermosetting resin systems usually have a viscosity at or below the critical 100 Nsm$^{-2}$, so that little or no shearing occurs. On the other hand, a considerable amount of shear orientation can occur in the injection molding of fiber filled thermoplastics. Takano[28] has reported improvements in orientability when the matrix viscosity in a fiber suspension is sufficiently high for some level of shear to occur in elongational flows. However, Modlen[32] has shown that the concurrent shear in an elongational flow may have a comparatively small effect on the final fiber orientation distribution. An additional effect in the injection molding of thermoplastics is that alignment would also be favored by the build-up of a frozen skin of shear-aligned material at the cold wall during flow. Resin–fiber separation also decreases with increasing resin viscosity.

Fisa and Utracki[33] have observed that mica platelets dispersed in polypropylene will align in the plane of the channel walls during extrusion through long dies. In a 40:1 $l/d$ rod die, a slow extrusion (1 s$^{-1}$) produced a concentric pattern. The orientation was more regular in the immediate vicinity of the wall than nearer the center. It also became disturbed as the mica concentration was raised, probably as the result of increased difficulty in packing, causing a folded pattern.

These results are somewhat contrary to earlier observations[34] showing only an incomplete orientation of the mica flakes parallel to the flow in a shorter (10:1) die. The difference could be attributed to the lack of shear, which in flake composites serves to refine the degree of parallel orientation. On the other hand, a decrease in the parallel orientation of short reinforcing fibers with increasing flow distance from the point of convergence is usually observed. Thus, shear alone, in the case of fiber reinforcement, is a disruptive force. The differences in the behavior between flake and fiber composites probably relate to the relative ease of packing dictated by the dimensionality of the particle.

Crowson et al.[18] report studies on fiber orientation in short (1/2 and 10 mm) glass fiber reinforced polypropylene undergoing converging, diverging, and shearing flows. By using a contact radiography technique, they found that the flow enters a capillary die through a very small cone angle from the reservoir; a high degree of fiber orientation parallel to the flow obtains as the entrance is approached, but much of this is lost during the shear flow through the tube. This behavior, along with the highly transverse orientation produced by a divergent flow, exactly parallels that found in epoxy BMC materials.[35] The orientation behavior in thermoplastic and thermosetting fiber composites appears to be identical in extent as well as in type.

### 6.3.2. Orientation in Reinforced Polymers During Processing

In processing operations, such as extrusion or injection molding, the hydrodynamic forces generated by the flow always produce some degree of fiber orientation. The purpose of the tool design is to establish certain types of flow fields which will cause the fibers to orient in the desired direction for the intended application. The extent, or degree, of fiber orientation depends on the intensity of the flow field and the response characteristics of the fibers to the flow.

As explained earlier, the highly pseudoplastic nature of composite flow resulting from the interactions between long fibers of high aspect ratio causes a blunted velocity profile in channels of constant cross-section. Thus, concentrated fiber suspensions do not orient in shear, but are highly sensitive to elongational flows. The fibers tend to align parallel to the direction of stretching or normal to the streamlines when compacted.

These normal deformations are generated by accelerating or decelerating the flow through changes in the cross-sectional area of the channel. The rate of fiber rotation is proportional to the size of the normal velocity gradient, and the direction of turning is determined by the sign of the velocity gradient. In accelerating flow fields, the velocity increases due to a decrease in the cross-sectional area of the channel in the flow direction, so that the velocity gradient is positive and the fibers rotate to make a smaller angle with the flow streamline on which they are travelling. The opposite occurs in a decelerating flow.

The orientation changes in the tool only if there is a change in cross-sectional area. Any shape which does not change the cross-sectional area between the channel inlet and outlet will result in no change to the fiber orientation. A higher alignment is produced as the difference between the cross-sectional area of the channel exit and that of the entrance section is increased. The area ratio is the most important variable influencing fiber orientation. Orientation in molding systems is probably determined more by the kinematics of the flow than by the hydrodynamic forces acting on individual particles, except as they influence the bulk constitutive properties of the suspension.

Orientation measurements can thus be correlated with computed and observed flow patterns in the mold. In one model of the orientation process[35] that applies to a converging axisymmetric section such as is shown in Fig. 6.8, shear is neglected for the highly concentrated fiber suspension considered, so that

$$v_1 = Q/A \qquad (6.8)$$

where $Q$ = volumetric flow rate, $A$ = cross-sectional flow area and

$$v_2 = v_3 = 0 \qquad (6.9)$$

The model of a structured continuum is well suited to describe orientation effects in concentrated suspensions of rods where inter-particle interactions are high. Mathematical theories of anisotropic fluids based on a continuum approach agree with experimental observations on the rotation of a suspended particle. Thus, in the resulting simple extension, the fiber orientation can be represented by the equations for a single ellipsoid[36]

$$d\theta_1/dt = \tfrac{3}{4}\lambda\dot{\varepsilon}\sin 2\theta_1 \qquad (6.10)$$

$$d\phi_1/dt = 0 \qquad (6.11)$$

By substituting the substantive time derivative and invoking the chain rule,

$$d\theta_1/dt = v_1\, d\theta_1/dx_1 = v_1(dA/dx_1)(d\theta_1/dA) \qquad (6.12)$$

In addition,

$$\dot{\varepsilon} = dv_1/dx_1 = (dv_1/dA)(dA/dx_1) \qquad (6.13)$$

but, since $Q$ is constant, eqn (6.8) gives

$$dv_1/dA = -v_1/A \qquad (6.14)$$

Therefore, eqn (6.10) becomes

$$d\theta_1/dA = -(3\lambda/4A)\sin 2\theta_1 \qquad (6.15)$$

which integrates to

$$\tan\theta_1 = C_1 A^{3\lambda/2} \qquad (6.16)$$

More simply,

$$\phi_1 = C_2,$$

where $C_1$ and $C_2$ are constants.

Although this model may not be rigorous, it can be used to interpret orientation phenomena, even in thermoplastic systems. It would apply to any axisymmetric converging section of an injection mold, and to most extrusion dies which reduce in area from the extruder head to the outlet orifice. It predicts that, since $\lambda$ is positive, the fiber orientation aligns more closely with the flow direction, as indicated by the streamline on which the fiber is moving, as the flow area decreases. If the streamlines are axisymmetric, so will the orientation distribution be, and the angle $\theta_1$ will indicate the spread in the distribution about the axis direction. Theoretical analyses of fiber rotations in dilute solutions predict $\lambda$ to be near unity. However, in the highly concentrated suspensions of 20–50 vol. % of long fibers generally encountered in the composites industry, the fiber–fiber interactions cause it to be reduced. It is similarly reduced for reinforcing fibers of $<50 \, l/d$ whose length is insufficient to develop the necessary hydrodynamics for fiber rotation. Typical values range from 0·5 to 0·8.

In Table 6.5 are listed the variables which may be expected to influence the fiber orientation. The first three of these are geometric variables. Of these, the area ratio is the most important since that variable appears explicitly in the mathematical model for $\theta_1$. All of the other parameters listed in the table might be important through their possible effect on the orientability parameter, $\lambda$. Variables (4) and (5) are entirely process-controlled, but (6)–(8) are determined in part by the material formulation. Variables (9) and (10) are clearly material variables. It should be noted that the flow rate, $Q$, and matrix viscosity, $\eta$, are probably combined in their effect through a Reynolds number,

TABLE 6.5

Variables Affecting Fiber Orientation in Processing of Composites

| (1) | Area ratio, $A/A_0$. |
|------|------|
| (2) | Angle of flow. |
| (3) | Channel diameter/fiber length. |
| (4) | Injection rate. |
| (5) | Cavity pressure during flow. |
| (6) | State of fiber aggregation. |
| (7) | Matrix viscosity. |
| (8) | Void content in the molding compound during flow. |
| (9) | Fiber concentration. |
| (10) | Fiber length/fiber diameter (aspect ratio). |

indicating that they act through an inertial effect, which is generally not significant in polymer processing. The conditions for optimum fiber orientability are as follows,

(1) Fiber aspect ratio sufficiently high so that moments will act to cause fiber rotation, but small enough to minimize bending and tangling. For a 12 $\mu$m diameter fiber, 3 mm length is optimum.
(2) Low fiber concentration and low compaction during flow to provide high rotational mobility.

The orientability parameter, $\lambda$, is not affected by small changes in matrix viscosity. A slight dependence on elongation rate occurs when the suspension is not fully compacted during flow. The angle of the die, channel diameter, state of aggregation and flow rate have little effect on fiber orientability.

Another flow field of application in polymer processing occurs when one coordinate dimension remains constant, while compensating changes are made in the other two to keep continuity satisfied. An example is the extrusion dies depicted in Fig. 6.14. Fiber orientation in this simple shear or planar extensional flow is described similarly to eqns (6.10) and (6.11) by

$$d\phi_i/dt = -(\lambda/2)(\partial v_j/\partial x_j) \sin 2\phi_i \qquad (6.17)$$

$$d\theta_i/dt = (\lambda/4)(\partial v_j/\partial x_j) \sin 2\theta_i \cos 2\phi_i \qquad (6.18)$$

In steady flow $v_i = 0$ while $v_j = f(x_j)$ and $v_k = f(x_k)$ to satisfy continuity. Again, under the assumption of negligible shear so that $v_1 = Q/A$ and invoking similar arguments to those used above it can be easily shown that eqns (6.17) and (6.18) reduce to

$$d\phi_i/dA = (\lambda/2A) \sin 2\phi_i \qquad (6.19)$$

$$d\theta_i/dA = -(\lambda/4A) \cos 2\phi_i \sin 2\theta_i \qquad (6.20)$$

which integrate to

$$\tan \phi_i = C_3 A^\lambda \qquad (6.21)$$

$$\tan \theta_i = C_4/(\sin 2\phi_i)^{1/2} \qquad (6.22)$$

As before, the important orientation angle (here $\phi_i$) depends only on the area change, with the rate of change determined by magnitude of the positive orientability coefficient, $\lambda$.

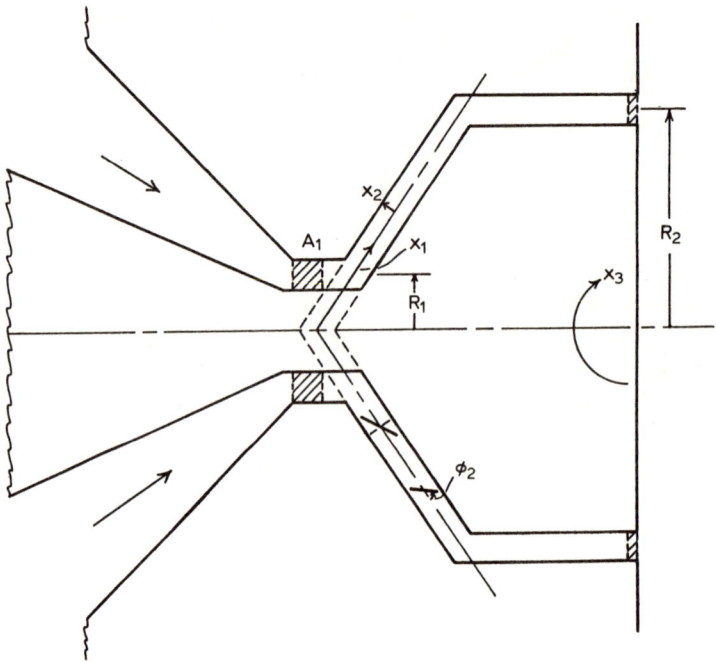

**Fig. 6.14.** Schematic drawing of an expanding mandrel die for producing circumferential orientation of reinforcing fibers in extruded pipe or hose (after Ref. 56).

The expanding mandrel die of Fig. 6.14 produces a circumferential fiber orientation because as the annular area, $A_1$, increases, $\tan \phi_2$, the angle measured in the plane of the flow, approaches $\pi$ radians, i.e. is directed in the circumferential direction.

## 6.4. INJECTION MOLDING

Fiber filled thermoplastics are commercially molded only on injection molding machines. The fiberglass is used to improve strength and modulus or to add dimensional stability and high temperature resistance. Thermoplastic molding materials may be purchased in different forms, depending upon the economics of the volume usage. The small molder can most profitably purchase the pellets of fiberglass and resin premixed to his desired fiber content. The next step is to buy a concentrate containing about 40 vol. % fiber, to be diluted with pure unfilled resin of the same type in the molding machine. Thus, by purchasing only two feeds (namely, concentrate and pure resin) a molder can have at his disposal a broad range of composite concentrations. Finally, resin and chopped fiber can be fed separately to the molding machine and mixed during plasticization in the screw.

Injection molding is highly automated in that the charge is automatically fed into the machine, and is heated internally during its residence in the barrel. The entire process is carried out in a single step. Heating is attained partially through viscous energy dissipation in the flow process. Although injection molding was initially developed for thermoplastic resins, new designs allow the molding of filled thermosets, such as fiberglass reinforced polyester. The characteristic features of injection molding are that the charge is automatically fed from a hopper, usually into a reciprocating screw (although rams can also be used), and that the screw and barrel are maintained at a different temperature than the mold, from which they are separated by a nozzle. For thermoplastics, the barrel is hotter, but for thermosets the reverse is true to prevent premature cure before the charge is shot into the cavity.

The usual practice and machine design for injection molding of unfilled polymers is easily carried over into the molding of RTP with little modification. Larger gates and runners are required to accommodate the higher viscosity of the composites. Both fiber degradation and erosion of the channel walls are undesirable. These can be reduced by limiting screw speed and back pressure, avoiding chemical attack from binder decomposition products, using hardened alloy barrels and keeping the fibers surrounded by a film of resin. Somewhat higher barrel temperatures are also desirable.

The hydrodynamic forces of the flow through runner, gate, and mold cavity tend to orient the short fibers into orientation patterns which depend upon the geometry and processing conditions as well as on the material constitution. The resulting fiber orientation, which is inevitable in any flow process unless the fibers are so short that they contribute no strength reinforcement, can be either beneficial or detrimental, depending upon the degree of control that is exercised to keep fiber alignment in the direction of stress in the molded part.

Figure 6.6 shows a rod in which longitudinal orientation (in accordance with Section 6.3.2) was caused by the converging flow at its entrance and the shear flow along its length. In the diverging flow that occurs instead following a small gate in a bar cavity, the fibers become oriented perpendicular to the streamlines of flow. The resulting orientation is primarily across the bar, but when the change in the cross-sectional area is abrupt the streamlines extend as spokes of a wheel from the point of gate attachment, giving rise to a component of longitudinal orientation near the wall. This may be subsequently enhanced by interactions with the wall. The change in orientation from the gate to the cavity occurs over a region of about 12 mm. The overall structure of the bar, then, is a transversely oriented core surrounded by a longitudinally oriented skin. There is no evidence, however, of large-scale fiber–resin separation that would be manifested as a variation in fiber content of the molding. No variation in fiber loading occurs along the length of the bar.

Tensile properties of the rods of Fig. 6.6 flow-molded from a premix epoxy molding compound through a converging flow field increase as the flow channel is reduced in area, to improve the degree of fiber alignment.[35] However, maximum possible alignment is limited to an average polar angle of about 20° to the flow axis. Flow disturbances, due to the intermittent nature of the flow, disrupt the orientation patterns when the cavity is too small, so that fiber orientation and the mechanical properties of the rod are not predictable. The maximum attainable properties from this composite comprising 40 vol. % fiberglass in a structural epoxy matrix are a Young's modulus of 25 GPa and a tensile strength of 185 MPa.

Schmidt[37] has experimentally studied and mathematically analyzed the flow fields developed in filling injection mold cavities and related them to fiber orientation patterns obtained. He describes a planar extension in the vicinity of the gate followed by a shear flow coupled with a fountain effect at the melt front as it travels down the cavity. The mold geometry is found to have the greatest effect, although mold temperature can change the development of the flow fields. The flow patterns are similar with unfilled, 19 wt % bead filled, and 16 wt % fiberglass filled polypropylene. The usual fiber orientation pattern is developed — namely, longitudinal orientation by stretching in the fountain flow at the melt front and shear at the walls with a transverse core resulting from the diverging entrance flow from the gate.

Recent studies on the flow and molding behavior of fiber reinforced polymers have been published by Folgar and Tucker,[24] Chan *et al.*[38] and Xavier *et al.*[39] Utracki[40] presents an excellent description of the issues to be resolved in the processing of fiber reinforced polymers in the light of their peculiar rheology. A recent review by Hegler[41] and the comprehensive monograph by Folkes[42] both cover in detail the multilayered morphological structure produced in the injection molding of fiber reinforced thermoplastics. As many as five distinct orientation layers from the centerline to the wall result from a combination of the entrance flow, the fountain flow effect at the melt front and shear along the frozen layer at the wall.

However, the state-of-the-art has not yet advanced to the stage of controlling orientations to optimize properties. Instead, fiber orientation is diminished by the use of very short fibers ($l/d < 40$), and the consequent loss of mechanical properties (tensile strength, at least) may be partially compensated by the attainment of a very high degree of dispersion of the fibers in the molding compound. There is then no worry about fiber damage in the screw, runners or gate, since even under repeated moldings a lower limit of $l/d$ of about 30 is attained, as shown in Table 6.6.

TABLE 6.6
Effect of Repeated Remolding in Producing a Limiting Aspect Ratio in Brittle Reinforcing Fibers

| Composite material | No. of passes through injection molding machine | Volume-median aspect ratio | Composite tensile strength (MPa) |
|---|---|---|---|
| Graphite fiber in polycarbonate | 0 | 1 060 | — |
| | 1 | 108 | 64·1 |
| | 2 | 42 | 72·4 |
| | 3 | 32 | 64·8 |
| Graphite fiber in nylon 66 | 1 | 51 | 124·8 |
| | 2 | 34 | 124·1 |
| | 3 | 34 | 111·0 |

In the flow of medium viscosity (10–100 Nsm$^{-2}$) fiber slurries through gates, it has been found[43] that gate size in the range of 3–12 mm does not have a strong effect on the length distribution of nominally 3 mm fibers. However, when one dimension is reduced below 0·8 mm, more extensive damage does result. Most damage occurs in the converging flow field found at die and gate entrances, where the suspension is subjected to both crushing and high elongational stresses. Most of the pressure drop also occurs at this point, though this is an extensional loss and does not necessarily reflect fiber damage. However, fiber damage is not influenced by the size of the convergence angle. Very little fiber degradation occurs during flow through uniform channels, even though some shearing of the fibers past one another may occur when the resin viscosity exceeds $100 \, \text{Nsm}^{-2}$.

With thermosetting matrices, tensile strengths are substantially lower when the fiber length is short (comparable with injection molded RTP). For example, in a fiberglass/epoxy composite, the tensile strength drops from 205 to 55 MPa as the fiber length is reduced from 6 mm to 1·6 mm or less. In such systems for which the matrix has a low ($< 1·0\%$) elongation to failure, long fibers are required and there is a greater need to control fiber orientation. That is not so much the case with thermoplastic matrices, especially nylon, as the data of Table 6.7 suggest.

TABLE 6.7

Comparison of Tensile Properties for Long and Short Fiberglass Reinforced Nylon[a] in Screw and Ram Injection Molders

| Molding machine type | Fiber length in molding compound (mm) | Tensile strength (MPa) | Young's modulus (GPa) | Ultimate elongation (%) |
|---|---|---|---|---|
| Screw | 9 | 214 | 13·8 | 2·1 |
| Screw | 0·25 | 166 | 11·0 | 2·4 |
| Ram | 9 | 221 | 15·2 | 1·6 |
| Ram | 0·25 | 159 | 10·3 | 2·1 |

[a] 23 vol. % fiberglass in nylon 66.

There are some thermoplastic molding compounds containing longer fibers on the market (fiber length about 9 mm), and those in fact do give high stiffness and tensile strength when the fiber alignment is properly controlled and caused to lie in the direction of maximum stress in the molding. The improper use of these compounds, however, may result in the appearance of cracks, low mechanical performance, surface roughness and poor reproducibility. Table 6.7 lists the relative properties of ASTM tensile bars molded from long and short fiber molding compounds using both a reciprocating screw and a ram injection machine. The better preservation of fiber length in the ram molder leads to higher values of strength and modulus since the flow into the neck section of these bars produces some orientation parallel to the test length.

A good discussion of the effects of the composite structural parameters of fiber length, dispersion, concentration and wet-out on the mechanical properties of injection moldings is given by Egbers.[44] He points out the trade-off between good dispersion (and wet-out) provided by precompounded molding pellets that is necessary for good surface and the higher impact strength (due to longer fibers) obtainable from a dry blend of fiber and resin. The critical parameter is the degree of shearing imparted both in the compound preparation and in the molding operation itself. Important molding variables affecting mixing are check valve design, back pressure, screw speed, melt and mold temperature, gate size and fill time.

Crowson *et al.*[18] studied the effects of temperature, fiber length, and fiber concentration on the rheology of fiber-filled melts (polypropylene and nylon 66). At high shear rate with 30 wt % fiberglass, the power-law flow index approaches zero, indicating that the material is flowing as a plug. At very low shear rates, on the other hand, Bright *et al.*[45] found the velocity profile to be nearly parabolic, indicating Newtonian flow. Thus, it is not unexpected that the fibers increase the viscosity to a greater extent at low shear rates (when shear is widespread and fiber orientation is low) than at high rates. Fiber length influences viscosity in a similar manner to fiber concentration.

In addition, it follows that the cores of end-gated bars are strongly aligned in the transverse direction at high injection speed, but have a more longitudinal orientation when the cavity is filled slowly. Both the general pattern and the effects of rate agree with those found for thermoset composites by Goettler.[14]

Darlington and coworkers[46, 47] have also studied fiber patterns and related them to mechanical anisotropy in fiber reinforced polypropylene. A high level of transverse orientation was measured in an edge-gated disk that produces a high expansion in the flow area as the melt enters the cavity.[48] The transverse core was surrounded by random surface layers, and this orientation was not affected by length variation in the fibers studied.

Cogswell *et al.*[49] cite the earlier data of Stephenson *et al.*[50] on disks molded from fiberglass reinforced polypropylene and nylon 66 to show predominant longitudinal or transverse fiber orientation depending on position in the mold, gate geometry, polymer viscosity and molding conditions. (Parallel alignment predominates near the surface and transverse in the core, in agreement with earlier results.) Thus, there is a difficulty in obtaining representative test pieces with composites. In order to estimate the general orientation pattern that could result from complex mold flows, Owen *et al.*[51] propose an approximate method for analyzing the geometric factors.

Other effects on molding that are directly attributable to the inclusion of reinforcing fibers are warpage and weak weld lines. The former results from the greatly reduced shrinkage parallel to the direction of fiber orientation when the fiber pattern is not balanced. It can be reduced without a substantial loss in shrinkage by substituting a particulate filler for a part of the fiber component. Since the fibers do not protrude across the advancing melt front, due both to the diverging entrance flow and to the fountain flow at the front, weld lines tend to be substantially weaker than other parts of the composite. The best remedy is to position the gating so that the weld orientation is not critical. It would be helpful to increase injection speed to maintain temperature at the weld. Reducing fiber content may benefit uniformity, but would generally weaken the part.

## 6.5. EXTRUSION

### 6.5.1. Issues in Extrusion of Short Fiber Composites

The usage of short fiber and flake reinforcements in extrusion operations is limited by the weak structure of the reinforced thermoplastic melt as it emerges from the die. At high fiber loadings, particularly of long fibers, the extrudate tears easily under line tension and by shear from the die walls at the exit.[52] The particularly low shear strength of nearly aligned short fiber composites causes even tensile fractures to occur in a shear mode, as shown by Fig. 6.15. Exposed surfaces show weak shear planes propagated along the fiber interfaces.

Under less extreme conditions, the surface remains characteristically rough. The more viscoelastic rubber composites are more easily extruded, though problems of surface quality, fiber orientation and uneven flow still exist.

Turning the fibers transverse to the flow (see below) improves the surface while generating a high crosswise stiffening that may be useful for certain applications. When it is desirable to maintain an axial fiber configuration, some beneficial refinements to the die configuration and

**Fig. 6.15.** Fracture surface of extruded rod reinforced with cellulose fiber in a nearly longitudinal orientation.

processing conditions have been suggested[53] for an improved surface smoothness.

The effect of fiber length on die swell is interesting in that, if the die is very short or the fibers very long (more than twice the die diameter), the extrudate is highly swollen, contains protruding fibers and may be foam-like. A restricted die swell occurs with short fibers (less than twice the die diameter) in long dies.[18] Cole *et al.*[54] have described a foam extrusion process utilizing the die swell from a zero land die. The complete absence of die swell that occurs with high fiber reinforcement in long dies simplifies die design.

## 6.5.2. Fiber Reorientation in Extrusion Dies

In some important cases, the reinforcing fibers should lie transverse to the flow direction. One commercial application is the circumferential reinforcement of an extruded tube (pipe or hose) to counteract the higher hoop stress generated in the tube wall from a contained pressure.

A rotating die can be used to produce circumferential fiber orientation in fiberglass reinforced thermoplastic pipe through shearing effects generated at the wall. On the other hand, Goettler *et al.*[55] found that a stationary expanding die[56] can attain the same effect very efficiently when the fiber reinforcement is sufficiently high to induce some measure of plug flow. This technology, which is well suited to fiber reinforced rubber[55] and plasticized PVC,[57] has been extended to encompass automatic shaping of the extrudate at the die.[58] By offsetting the inner and outer portions of the hose die in a controlled sequence, a reproducible curvature can be imparted to the hose without sacrificing performance.

It has been shown that fibers flowing as a suspension in a matrix vehicle respond to elongational flow fields according to eqns (6.10)–(6.22). The combination of shear forces with the overall reduction in cross-sectional area occurring between the head of the extruder and the die orifice of a conventional extrusion die would cause the fibers to become aligned parallel to the axis of the extrudate. A new design comprises a restriction of the flow at some intermediate point within the die, followed by a conical expansion to form the dimensions of the extruded product. This 'conically expanding die' imparts a hoop-wise transverse orientation pattern to the fibers, as shown in Fig. 6.14.

In these conically expanding dies, the channel thickness is kept constant and the material is stretched circumferentially around the mandrel as it flows through the die. The resulting degree of hoop orientation increases as the area expansion ratio increases. When the channel thickness is constant, as here, the cross-sectional area expansion ratio becomes numerically equal to the ratio of the radius at the outlet to that at the constriction, $R_2/R_1$.

In expanding extrusion die geometries which induce normal velocity gradients in the melt flow, the orientation of the fibers within the transverse plane is dependent upon the relative magnitudes of the gradients $dv_2/dx_2$ and $dv_3/dx_3$. The degree of circumferential fiber orientation is proportional to a power of the ratio of outlet and inlet diameters in the die, whose channel expands along a conical contour. If the channel thickness also increased, either alone or in conjunction with the conical expansion, the radial fiber orientation would be enhanced according to the same function of the ratio of outlet to inlet channel thicknesses.

Instead of measuring fiber angles, the degree of fiber orientation can be interpreted in terms of the mechanical anisotropy based on directional measurements of:

(a) Young's modulus;
(b) growth under hydrostatic pressure;
(c) swelling after imbibing of a solvent.

The circumferential Young's modulus, for example, increases with the diameter expansion in the die. The magnitude of this effect is far greater than would have been anticipated from the state-of-the-art on fiber reinforced plastics extrusion. In PVC reinforced with cellulose fiber, the circumferential modulus can become five times that in the axial direction.[57] With fiberglass in polypropylene, a more random orientation is attributed to the fracture of the glass to a small aspect ratio.[53]

In a hose, the expanding die can impart equal hoop and axial growth under pressure, as well as an enhanced burst strength that is related to the diameter expansion in the die and to the degree of fiber reinforcement in the stock. Solvent swell data indicate that circumferential reinforcement is related to the radius expansion in the die whereas the degree of radial reinforcement is a function of the channel width, as explained above. Axial reinforcement is inversely related to the overall area expansion.

The stresses in straight lengths of extruded composite tubes under an internal pressure loading may be estimated from the formulae for thin-walled cylindrical pressure vessels:

$$\sigma_H = DP/2t \qquad (6.23)$$

$$\sigma_A = DP/4t \qquad (6.24)$$

where $\sigma_H$, $\sigma_A$ are stress in hoop and axial directions, $D$ is diameter, $t$ is wall thickness and $P$ is pressure.

In designing against rupture, a high tensile strength in the circumferential direction is necessary to resist the high wall hoop stress. Since the directional variation in tensile strength with fiber angle is much less than for Young's modulus, it is not likely that the optimum 2:1 strength ratio in the hoop/axial directions, as indicated by eqns (6.23) and (6.24), will be exceeded within the range of practical expansion ratios

$(R_2/R_1 < 6)$. In the selection of the expansion ratio for the die, it should be recalled that the burst strength, $P_u$, depends on both the properties of the composite material comprising the tube and the tube geometry. In turn, the material properties of the extrudate depend on its fiber orientation pattern and on the properties of a highly aligned sample of the same material. It has been found that the burst strength correlates linearly with $\sigma_0$, the ultimate tensile strength of a uniaxial fiber array at an angle of $0°$ to the fiber direction, giving

$$P_u = 2K(t/D)\sigma_0 \qquad (6.25)$$

This equation is simply a rearrangement of eqn (6.23) in which the hoop strength is replaced by its equivalent: the product of the compound tensile strength and an orientation function, $K$. $K$ increases as the radius expansion ratio $R_2/R_1$ increases in expanding mandrel dies.

It has been found that eqn (6.25) describes the main parameter dependence of the burst strength. All other parameters exert only secondary effects. There is no variation of the dimensionless group $DP_u/2t\sigma_0$ with either fiber concentration or Young's modulus of the aligned stock. There is no significant effect of varying the channel thickness, the land length, surface condition or divergence angle, or from the extrusion conditions.

### 6.5.3. Prepregging

Besides the direct extrusion of continuous profiles, extrusion can be used as a method of prepreg manufacture with either thermoplastic or thermosetting matrices. By the proper design of the converging channel of the extruder die, orientation distributions in which the average fiber deviation from the axis is $20°$ have been produced. This is about the same angle as that obtained in flow molding. Prepregged strands of oriented discontinuous glass, boron, and graphite fibers in epoxy and polycarbonate resin prepared in this way have been compression molded into test specimens whose specific tensile properties often exceed those of aluminum. However, the tensile strengths of such moldings (with an average fiber angle of $20°$) are no better than for a two-dimensionally random composite, and are only half that obtainable with highly aligned continuous fibers at the same loading. Data are presented in Table 6.2.

Again, strength is limited by cleavage-type breaks around fiber bundles. These have been observed extensively in all discontinuous-fiber flow-processed composites in which the fibers are sufficiently long.

## 6.6. SPECIALTY PROCESSING TECHNIQUES

### 6.6.1. Prepregging

Several other processes have been developed for pre-aligning fibers before they are infiltrated with resin. These employ the stretching flow through a tapered tube or the shear developed in dilute suspensions flowing through uniform channels to align the fibers into the direction of flow. The advantage over high shear melt processing is the greater control over fiber alignment that is possible.

The effluent, in the form of a strand or mat of intermeshed but oriented fibers, has little integrity. It is carefully laid down or wound onto a spindle or drum, washed, and then laid up for resin impregnation. A scheme developed in England utilizes a suspension of the short fibers in glycerin.[59] Wide mats can be produced by elongating the material as it flows through a slit in a moving trough. Washing and impregnation follow.

Whilst thermosetting resins can be applied as a prepolymer, thermoplastics are usually coated from the melt or solution. The latter would produce better wet-out, but precautions must be taken to insure solvent removal. Tensile properties competitive with thermosetting composites are possible.

By using a refined extrusion technique along with a reciprocating filter bed, Salariya and Pittman[60] were able to align a sheet of single carbon fibers to 90% within 7° of the flow axis. They developed methods to predict the aligning power of a converging channel, the velocity profile in the sheet and the extensional viscosity of the fiber suspension. The most perfect fiber alignment in short-fiber composites can only be obtained by tedious methods of hand lay-up of well-oriented highly elongated grains of resin coated fibers.[28] The composites made by hand laying these elongated grains not only excel in precision of alignment (± 7°), but also are unique in simultaneously affording high degrees of fiber wet-out, dispersion, and end overlapping. The values of tensile strength and modulus obtained by this method can be attributed to the large percentage (about 40%) of the fibers that are fractured in a tensile failure, compared, for example, with almost none in conventional injection moldings.

## 6.6.2. Monomer Casting and RRIM

The best resin–reinforcement contact in a thermoplastic system can be obtained when the two are combined before polymerizing the resin.[61] Then, viscosity of the monomer is in the $10^{-3}$ Nsm$^{-2}$ range. In addition, void elimination is enhanced and high fiber loadings with less fiber damage are possible. For example, composites prepared by polymerizing $\varepsilon$-caprolactam (nylon 6) around a highly aligned array of glass fibers display high tensile strength and elongation to failure in the transverse direction. The tensile strength of 83 MPa, equal to that of the matrix, and an elongation of 2·0%, achievable at 30 vol. % fiber content, are unknown for other non-elastomeric polymer systems. In contrast, a typical brittle epoxy reinforced with fibers transverse to the stress direction breaks at 0·5% strain and about 55 MPa tensile stress. With such a material, classical design limits for laminated structures are surpassed. The superior transverse tensile properties of the nylon composite are attributed to several factors: the excellent adhesion of the high polarity nylon to the glass surface, the excellent wet-out of the fibers, and the ductility of the nylon, including its ability to cold draw. A fibrillar structure was observed in the tensile fracture surfaces by scanning electron microscopy. It was accompanied by the formation of microvoids.

Chopped glass fibers up to 6 mm in length are now being used in reinforced reaction injection molding of polyurethanes. Milled glass at 25% concentration increases the prepolymer viscosity from 0·2 to 0·6–1·0 Nsm$^{-2}$. To keep the viscosity from being boosted further, the concentration of the longer chopped fiber is usually limited to about 3%. However, equivalent mechanical properties result. The strong orientation produced by the chopped fibers is observed parallel to the mold filling flow, changing properties by a factor of 2, but randomization through prudent gate placement is usually attempted to limit warpage.

### 6.6.3. Stamping

Thermoplastic sheets randomly reinforced with fiberglass are formed by modified metal stamping presses. Productivity is high, and capital investment less than for injection molding. Indeed, the latest developments in fiber reinforcement of thermoplastics utilize continuous or chopped cord laminated in a random pattern between nylon or polyphenylene sulfide sheets, which can be formed by stamping. Other thermoplastic matrices have also been used.

Planar isotropic composites made from *in situ* polymerized nylon 6 and 13 mm glass mat display a tensile strength in excess of 200 MPa at less than 30 vol. % fiber concentration.[61] Ultimate elongations are in the range 2·0–2·5%, compared with less than 2·0% for epoxy and the aforementioned thermoplastic sheet composites. This indicates a high level of matrix performance for the nylon, since failure tends to be fiber controlled. The elastic modulus is independent of matrix type since it is a low elongation property, and attains a level of 12·4 GPa at 30 vol. % fiber.

The effects of extruded sheet stretching on fiber orientation and fiber length distributions in short fiber reinforced mats for deep drawing and hot stamping operations are described by Nicolais *et al.*[62] As the draw ratio is increased, the fibers rotate into the stretch direction according to a model based on the overall sample deformation, and are simultaneously fractured when the stretching stress exceeds the fiber strength. The simple models (shear lag stress analysis and a special case of the rotational orientation model with $\lambda = 1$) are adequate for the low (< 1 %) fiber concentrations studied.

## 6.7. SUMMARY

The knowledge of composite rheology and the conditions for controlling reinforcement size and orientation is essential if full advantage is to be taken of the opportunities for fabrication offered by flow processing. It must involve an interaction of tool geometry, processing conditions, and composition of the composite. The resulting quantitative predictions can then be applied to the design of complicated shapes. It is by facilitating this final step that the more basic approach to the study of composite morphology in flow processing surpasses in utility the collected qualitative observations of the earlier literature.

Rheological studies on suspensions must deal effectively with particle interactions in concentrated non-Newtonian viscoelastic media if they are to be applied to the interpretation of composite behavior. Mathematical analyses of mold filling now under development should similarly allow for the anomalous flow properties of these concentrated suspensions, since it is found that the fluid mechanics is strongly coupled to the anisotropic rheology. Finally, to predict part performance, the theories of anisotropic mechanical properties in composite materials need to accommodate complicated generalized orientation distributions. Clearly, the overall design problem involves an integration of part design with the composite material properties, which are themselves defined by the geometrical changes occurring during the fabrication step.

## REFERENCES

1.  Mandell, J. F., Darwish, A. Y. and McGarry, F. J., 'Fracture testing of injection molded glass and carbon fiber reinforced thermoplastics', *ASTM Symp. on Test Methods and Design Allowables for Fibrous Composites, Dearborn, MI,* 1979.
2.  McNally, D. L., Freed, W. T., Shaner, J. R. and Sell, J. W., *Polym. Eng. Sci.,* 1978, **18**, 396.
3.  Darlington, M. W. and Smith, G. R., *Polymer,* 1975, **16**, 459.
4.  Moskal, E. A., *Plastics Design and Processing,* 1977, **17**(1), 10.
5.  Stade, K., *Polym. Eng. Sci.,* 1977, **17**, 50.
6.  Lunt, J. M. and Shortall, J. B., *Plastics and Rubber: Processing,* 1979, **4**(3), 108.
7.  Hamed, P. and Coran, A. Y., in *Additives for Plastics,* Vol. I, R. B. Seymour (Ed.), Academic Press, New York, 1978, p. 29.
8.  Schlich, W. R., Hagan, R. S., Thomas, J. R., Thomas, D. P. and Musselman, K. A., *SPE J.,* Feb. 1968, **24**, 43.
9.  Maschmeyer, R. O. and Hill, C. T., *Trans. Soc. Rheol.,* 1977, **21**, 183, 195.
10. Milewski, J. V., *Plastics Compounding,* Nov./Dec. 1979, 17.
11. Katz, H. S. and Milewski, J. V., *Handbook of Fillers and Reinforcements for Plastics,* Van Nostrand–Reinhold, New York, 1978.
12. White, J. L. and Spruiell, J. E., *Polym. Eng. Sci.,* 1983, **23**, 247; 1981, **21**, 859.
13. Pipes, R. B., McCullough, R. L. and Taggert, D. G., *Polymer Composites,* 1982, **3**, 34.
14. Goettler, L. A., *25th Conference SPI Div. Reinforced Plastics and Composites,* Washington, DC, 1970, Section 14A.
15. Hochschild, R., *Materials Evaluation,* 1968, **26**(1), 35A.
16. Botsco, R. J., *Plastics Design and Processing,* Nov./Dec. 1968.
17. Darlington, M. W. and McGinley, P. L., *J. Mater. Sci.,* 1975, **10**, 906.
18. Crowson, R. J., Folkes, M. J. and Bright, P. F., *Polym. Eng. Sci.,* 1980, **20**, 925; Crowson, R. J. and Folkes, M. J., *Polym. Eng. Sci.,* 1980, **20**, 934.
19. Schierding, R. G., *J. Composite Mater.,* 1968, **2**(4), 448.
20. Menendez, H. and White, J. L., *Polym. Eng. Sci.,* 1984, **24**, 1051.
21. McGee, S. H. and McCullough, R. L., in *The Role of the Polymeric Matrix in the Processing and Structural Properties of Composite Materials,* J. C. Seferis and L. Nicolais (Eds), Plenum, New York, 1983; p. 425.
22. Ishai, O. and Lavengood, R. E., in *Composite Materials: Testing and Design,* ASTM STP 460, 1969, pp. 271–81.
23. Goettler, L. A. and Lavengood, R. E., *American Chemical Society Div. Organic Coatings and Plastics Chemistry Preprints,* 1971, **31**(1), 623.
24. Folgar, F. and Tucker, C. L., *J. Reinf. Plastics Composites,* 1984, **3**, 98.

25. Givler, R. C., Crochet, M. J. and Pipes, R. B., *J. Composite Mater.*, 1983, 17, 330.
26. Goldsmith, H. L. and Mason, S. G., in *Rheology*, F. R. Eirich (Ed.), Academic Press, New York, 1967, pp. 85–250.
27. Bell, J., *J. Composite Mater.*, 1969, 3, 244.
28. Takano, M., *Viscosity Effect on Flow Orientation of Short Fibers*, US Defense Documentation Center, Arlington, VA, Report No. AD-772563, 1973.
29. Harris, J. B. and Pittman, J. F. T., *Trans. Instn Chem. Engrs*, 1976, 54, 73.
30. Lee, W.-K. and George, H. H., *Polym. Eng. Sci.*, 1978, 18, 146.
31. McNally, D. L., *Polym. Plast. Technol. Eng.*, 1977, 8(2), 101.
32. Modlen, G. F., *J. Mater. Sci.*, 1969, 4, 283.
33. Fisa, B. and Utracki, L. A., *Polymer Composites*, 1984, 5, 36.
34. Okuno, K. and Woodhams, R. T., *Polym. Eng. Sci.*, 1975, 15, 308.
35. Goettler, L. A., *Polymer Composites*, 1984, 5, 60.
36. Takserman-Krozer, R. and Ziabicki, A., *J. Polym. Sci., A1*, 1963, 491.
37. Schmidt, L. R., *Polym. Eng. Sci.*, 1974, 14, 797.
38. Chan, W. W., Charrier, J. and Vadnais, P., *Polymer Composites*, 1983, 4, 9.
39. Xavier, S. F., Tyagi, D. and Misra, A., *Polymer Composites*, 1982, 3, 88.
40. Utracki, L. A., *Rubber Chem. Technol.*, 1984, 57, 507
41. Hegler, R. P., *Kunststoffe*, 1984, 74(5), 271.
42. Folkes, M. J., *Short Fibre Reinforced Thermoplastics*, Research Studies Press, Chichester, 1982.
43. Goettler, L. A., *Modern Plastics*, April 1970, 48, 140.
44. Egbers, R. G., *Plastics World*, Oct. 1979, 66.
45. Bright, P. F., Crowson, R. J. and Folkes, M. J., *J. Mater. Sci.*, 1978, 13, 2497.
46. Darlington, M. W., McGinley, P. L. and Smith, G. R., *Plastics and Rubber: Materials and Applications*, May 1977, 51.
47. Darlington, M. W., McGinley, P. L. and Smith, G. R., *J. Mater. Sci.*, 1976, 11, 877.
48. Darlington, M. W., Gladwell, B. K. and Smith, G. R., *Polymer*, 1977, 18, 1269.
49. Cogswell, F. N., Cole, E. A. and Turner, S., *J. Elast. Plast.*, 1979, 11, 171.
50. Stephenson, R. C., Turner, S. and Whale, M., *Society of Plastics Engineers Annual Technical Conference*, Montreal, Canada, 1977, p. 347.
51. Owen, M. J., Thomas, D. H. and Found, M. S., *Modern Plastics*, 1978, 55(6), 61.
52. Goettler, L. A., Sezna, J. A. and DiMauro, P. J., *Rubber World*, Oct. 1982, 181(1), 33.
53. Goettler, L. A., in *The Role of the Polymer Matrix in the Processing and Structural Properties of Composite Materials*, J. C. Seferis and L. Nicolais (Eds), Plenum, New York, 1983, p. 289.
54. Cole, E. A., Cogswell, F. N., Huxtable, J. and Turner, S., *Polym. Eng. Sci.*, 1979, 19, 12.
55. Goettler, L. A., Leib, R. I. and Lambright, A. J., *Rubber Chem. Technol.*, 1979, 52, 838.
56. Goettler, L. A. and Lambright, A. J., to Monsanto Company, US Patent 4 056 591, Nov. 1977.

57. Goettler, L. A., *Polymer Composites,* 1983, **4**, 249.
58. Goettler, L. A., Lambright, A. J., Leib, R. I. and DiMauro, P. J., *Rubber Chem. Technol.,* 1981, **54**, 277.
59. Bagg, G. E. G., Evans, M. E. N. and Pryde, A. W. H., *Composites,* Dec. 1969, 97.
60. Salariya, A. K. and Pittman, J. F. T., *Polym. Eng. Sci.,* 1980, **20**, 787.
61. Goettler, L. A., 'Mechanical performance of various nylon 6 composites formed by *in-situ* polymerization of caprolactam', *American Chemical Society Polymer Preprints,* for Spring 1985 meeting.
62. Nicolais, L., Nicodemo, L., Masi, P. and DiBenedetto, A. T., *Polym. Eng. Sci.,* 1979, **19**, 1046.

# Chapter 7

# Procedures for Engineering Design with Short Fibre Reinforced Thermoplastics

## M. W. DARLINGTON

*Cranfield Institute of Technology, Bedford, UK*

and

## P. H. UPPERTON

*Du Pont (UK) Ltd, Hemel Hempstead, UK*

## 7.1. INTRODUCTION

For many engineering or structural applications, plastics, due to their relatively low moduli, often compare unfavourably with more traditional materials such as wood and metal. This limitation in stiffness has not, however, prevented plastics from being used in a wide range of load-bearing applications. A suitable choice of geometry and an efficient use of stiffening ribs enable components which exhibit adequate overall stiffness to be designed from these low modulus materials.

Although they can be used to replace the more traditional materials, the deformation behaviour of plastics leads to greater problems at the design stage. The major problem encountered is that the mechanical properties are dependent upon time under load and temperature. These two factors dominate any design procedures because the data used in any calculations must be appropriate to the service conditions.

Procedures for design for stiffness that allow for this complex behaviour are now well developed for unfilled thermoplastics. Progress has also been made in some aspects of design for strength. An appreciation of current approaches to data presentation and design with unfilled thermoplastics forms an essential background to the discussion of the special considerations and problems associated with short fibre reinforced thermoplastics. A brief survey of those aspects of the subject which are relevant to the subsequent discussion on short fibre reinforced thermoplastics is, therefore, presented in Section 7.2. Greater detail, covering a wider range of topics, may be found elsewhere.[1-3]

One of the main reasons for the addition of short fibres to a thermoplastic is to increase the stiffness of the base polymer and so extend its range of applications in load-bearing situations. However, the fibres introduce several additional factors which affect the mechanical properties of the material and hence further complicate the tasks of data generation, data selection and design. These factors are discussed in Section 7.3.

Cost considerations limit the volume of data which materials suppliers can be expected to generate. Uncertainty, due to a lack of knowledge or understanding of materials behaviour, leads to overdesign and hence uneconomic solutions. If short fibre reinforced thermoplastics are to be fully exploited, realistic design procedures must be available which do not place undue demands on data generation and which offer the designer an acceptable compromise between accuracy and ease of use. The main purpose of this chapter is to review progress in this area. Particular attention is paid to design for stiffness but consideration is also given to design for strength.

## 7.2. MECHANICAL PROPERTIES AND DESIGN PROCEDURES FOR UNFILLED THERMOPLASTICS

### 7.2.1. Deformation Behaviour

The deformation behaviour of plastics is classed as viscoelastic which, in the context of this chapter, can be taken to mean 'time-dependent elastic'. Data from tests in which specimens are deformed at constant extension rate are seldom suitable for design purposes. More relevant data are obtained from creep tests or stress relaxation tests. There is little to choose between the two methods although creep experiments have become the more popular.

In creep experiments, a constant force is applied to a specimen and the resulting strain, which increases with time under load, is monitored. If the force is removed, the strain decreases with increasing time and will often tend to zero (i.e. complete recovery) at long times (see Fig. 7.1). A creep modulus, $E(t)$, can be defined from this test:

$$E(t) = \frac{\sigma_0}{\varepsilon(t)}$$

where $\sigma_0$ is the constant applied stress and $\varepsilon(t)$ is the resultant time-dependent strain.

In stress relaxation experiments, a constant strain is imposed upon a specimen and the force required to maintain that strain is monitored. The idealised response is shown in Fig. 7.2. A stress relaxation modulus, $M(t)$, can be defined:

$$M(t) = \frac{\sigma(t)}{\varepsilon_0}$$

where $\varepsilon_0$ is the constant strain and $\sigma(t)$ is the resultant time-dependent stress.

It may be shown theoretically that the stress relaxation modulus will *not* be equal to the creep modulus for the same basic deformation mode. However, for many plastics, in many situations, the error incurred in assuming $M(t) = E(t)$ will be relatively small.

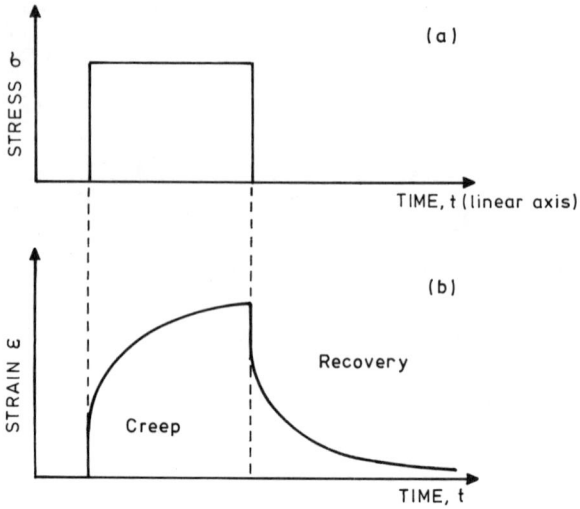

**Fig. 7.1.** Creep and recovery of a viscoelastic material. (a) Applied stress history; (b) strain response.

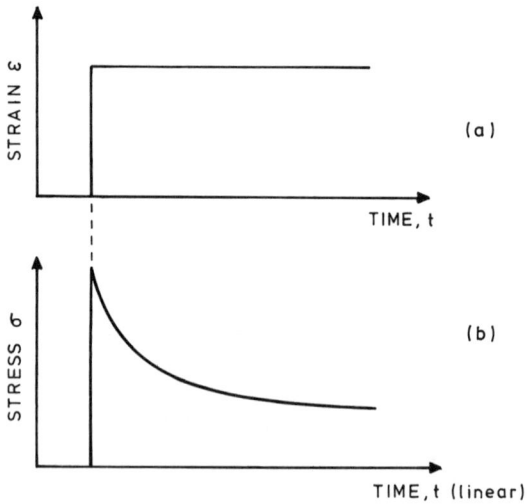

**Fig. 7.2.** Stress relaxation of a viscoelastic material. (a) Applied strain history; (b) stress response.

If two separate creep tests, at stresses $\sigma_1$ and $\sigma_2$, result in strains of $\varepsilon_1(t)$ and $\varepsilon_2(t)$ respectively, as a function of time then from test 1:

$$E_1(t) = \sigma_1/\varepsilon_1(t)$$

and from test 2

$$E_2(t) = \sigma_2/\varepsilon_2(t)$$

If $E_1(t) = E_2(t)$ for the time range studied, the material is said to be *linear viscoelastic* over the time and stress range, i.e. the creep modulus, $E(t)$, is independent of stress, $\sigma$.

However, for most thermoplastics, over their normal working stress or strain range, it is found that $E_1(t) \neq E_2(t)$ and the materials are said to be *non-linear viscoelastic*, i.e. the creep modulus, $E(t)$, depends on the magnitude of the applied stress.

From the two separate creep tests, if $\sigma_2 = 2\sigma_1$, then for a linear viscoelastic material $\varepsilon_2(t) = 2\varepsilon_1(t)$ for all values of time, $t$. For a non-linear viscoelastic material, we would usually find that $\varepsilon_2(t) > 2\varepsilon_1(t)$ and the ratio $\varepsilon_2(t)/\varepsilon_1(t)$ would itself change with time.

Similar considerations apply for the stress relaxation modulus, $M(t)$. For example, the material is non-linear viscoelastic if $M(t)$ depends on the magnitude of the applied strain.

The above pattern of behaviour is observed in all deformation modes (uniaxial tension/compression, shear and bulk.)

### 7.2.2. Presentation of Creep Data

Much of the available design information for plastics has been obtained from creep tests in uniaxial tension. The presentation of such data for design purposes is described in detail in British Standard BS4618.[4] A logarithmic time scale is used to cover the wide time range of interest. The curve shape shown in Fig. 7.1 then converts typically to a curve of the shape shown in Fig. 7.3.

Since the materials are usually non-linear viscoelastic, creep tests are generally carried out at several stress levels (at constant temperature) and the basic data presentation is a set of curves showing the variation of strain with log (elapsed time) at each stress level. A typical family of creep curves at one temperature is presented in Fig. 7.3.

There are several ways in which the stress–strain–time dependence can be displayed. Extraction of data for design purposes may be easier using one of these alternative methods of presentation. The most important cross-plots are illustrated in Fig. 7.4.

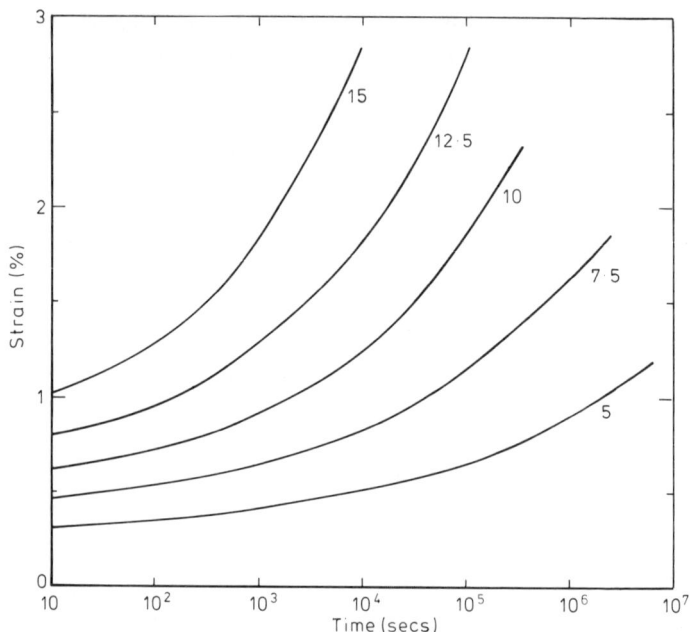

**Fig. 7.3.** Family of strain–log (time) creep curves for a range of applied stress levels (MPa). Polypropylene homopolymer tested in uniaxial tension at 23 °C.

By plotting the applied stress level against the resulting strain reached in a chosen time, an *isochronous* stress–strain cross-plot is produced. The non-linearity of the material response (i.e. lack of proportionality of stress and strain) is immediately apparent from such a plot.

Alternatively, a strain level can be selected and the stress plotted against the log of the time taken to reach the chosen strain level. This *isometric* stress–log (time) cross-plot is often taken as a first approximation to the stress relaxation behaviour of the material.

By dividing the time-dependent stress by the fixed strain level chosen for the isometric cross-plot, an isometric modulus-log (time) graph may be produced.

### 7.2.3. Factors Influencing the Creep Response
For a given sample of material, the major factor affecting the family of strain–log (time) curves will be the test temperature.[5] Some thermo-

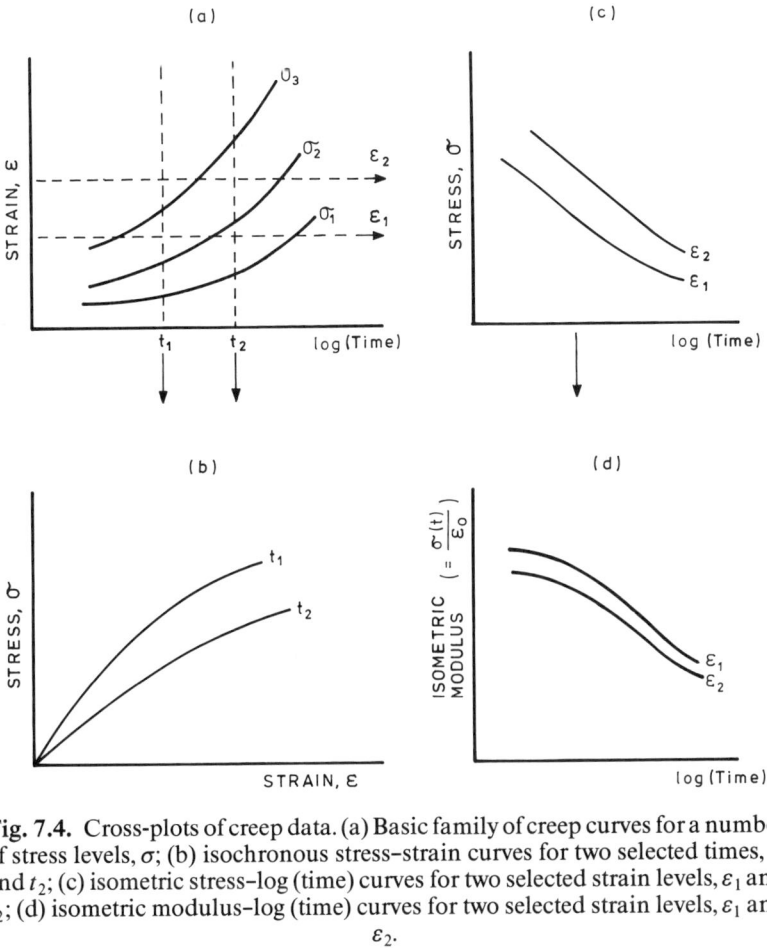

Fig. 7.4. Cross-plots of creep data. (a) Basic family of creep curves for a number of stress levels, $\sigma$; (b) isochronous stress–strain curves for two selected times, $t_1$ and $t_2$; (c) isometric stress–log (time) curves for two selected strain levels, $\varepsilon_1$ and $\varepsilon_2$; (d) isometric modulus–log (time) curves for two selected strain levels, $\varepsilon_1$ and $\varepsilon_2$.

plastics (e.g. nylons 6 and 66) absorb significant amounts of water, the equilibrium water content depending on the environmental conditions. For these materials, the family of strain–log time creep curves at any one temperature will itself be a function of the relative humidity.[6, 7]

The mechanical properties of plastics depend on the molecular state of the material when the test sample, or fabricated component, is tested. Since this molecular state depends on the processing history and the subsequent storage history, the overall pattern of creep behaviour referred to above (i.e. stress, temperature and relative humidity dependence) is a complex function of these histories.[5,7] It is probably this aspect of creep behaviour that leads to the greatest uncertainty in the selection of appropriate data for design purposes.

### 7.2.4. Principles of Design for Stiffness

The fundamentals of designing with plastics in situations where deformation under load is a major concern are based on the use of data applicable to the time-dependent behaviour of the material.

The most common approach to design for stiffness with plastics is that of the 'pseudo-elastic' design method. This encompasses a number of design procedures all based upon the one fundamental assumption:

'The stress distribution in an elastic structure is identical to the stress distribution in a linear viscoelastic structure of the same geometry and subjected to the same loading.'

Thus, design for stiffness with a viscoelastic material consists of identifying the appropriate structural idealisation, selecting the appropriate elastic formula and replacing the modulus and deformation functions by their respective time-dependent equivalents.

*Example*

For a cantilever beam rigidly supported at one end, with a force applied at the other end, the elastic formula for the deflection at the free end of the beam is given by eqn 7.1.

$$Y = \frac{FL^3}{3EI} \tag{7.1}$$

where $Y$ = deflection at free end of beam,
  $F$ = applied force,
  $L$ = length of beam,
  $I$ = second moment of area (moment of inertia) and
  $E$ = Young's modulus.

If the material is linear viscoelastic, the modulus, $E$, can be replaced by its time-dependent equivalent. If the force is constant, this will be the tensile creep modulus, $E(t)$. Equation 7.1 can therefore be rewritten to calculate the deflection due to the applied force, $F$, being applied for a specific period of time (eqn (7.2)).

$$Y(t) = \frac{FL^3}{3E(t)I} \tag{7.2}$$

Thus, if a constant force, $F$, is applied for a period of 10 min, the deflection can be calculated by inserting into eqn (7.2) the value of creep modulus appropriate to a creep time of 10 min. If the force is applied for a period of one year, then by inserting the creep modulus appropriate to a creep time of one year the deflection after one year can be calculated.

If the material behaves in a non-linear viscoelastic manner the value of $E(t)$ is dependent upon the stress level. To allow for this, a creep modulus needs to be used that is appropriate to the stress or strain conditions within the beam.

If plane sections at right angles to the neutral axis remain plane and the modulus of elasticity in tension equals the modulus of elasticity in compression, the stresses due to bending can be expressed by eqn (7.3).

$$\sigma = \frac{MC}{I} \tag{7.3}$$

where $\sigma$ = stress due to bending,
     $M$ = bending moment at appropriate section ($M = FL$, at fixed end for end-loaded cantilever beam),
     $I$ = second moment of area,
     $C$ = distance from neutral axis.

It is evident from eqn (7.3) that the maximum stress with an end-loaded cantilever beam occurs at the fixed end and at the outermost section of the beam. This is termed the maximum outer fibre stress and can be calculated by eqn (7.4).

$$\sigma_{m} = FLd/2I \tag{7.4}$$

where $\sigma_{m}$ = maximum outer fibre stress,
     $d$ = beam depth.

The equivalent outer fibre strain can be calculated by combining eqns (7.1) and (7.4).

As much of the material in the beam will be at a lower stress (and strain) than that at the outer fibre, the use of the maximum outer fibre stress in the selection of the appropriate creep curve for the calculation of creep modulus will ensure a 'safe' design.[8]

For simple problems of the above type, when the forces are constant, it is apparent that much of the effort of 'design' goes into the production and presentation of the creep data.

### 7.2.5. Strength Considerations

In common with stiffness, the strength properties of a given sample of a thermoplastic depend on the loading history and temperature. For the purposes of data presentation, loading histories have traditionally been grouped into three general categories: 'static', cyclic (fatigue behaviour) and high strain rate (impact behaviour).[9] It is not possible in this chapter to cover all these aspects of the strength properties and design for strength with thermoplastics. Only those aspects which are necessary to the limited discussion on the strength of short fibre reinforced thermoplastics in Section 7.6 are therefore discussed below.

Short-term strength data are usually derived from uniaxial tests carried out on a tensile testing machine operated at constant rate of crosshead separation. (For the purposes of this discussion, 'short-term' will be taken to mean times of at most a few minutes.) The basic load–extension data are often converted to a nominal stress–nominal strain curve, based on the original area of cross-section and original gauge length respectively. Tests at a chosen temperature and strain rate on different thermoplastics result in a variety of curve shapes,[9] some of which are shown in Fig. 7.5.

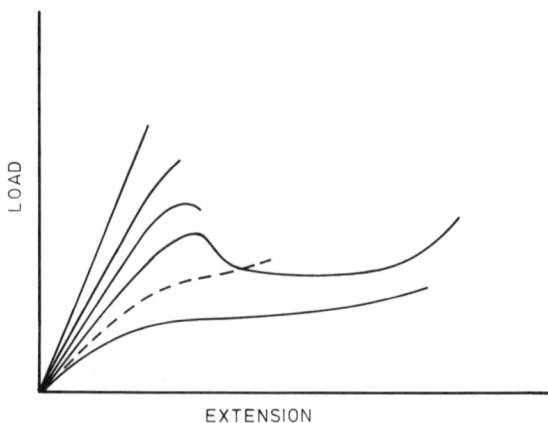

**Fig. 7.5.** Typical load–extension curves from tests at constant extension rate on thermoplastics.

Many materials show a peak in the stress–strain curve. This peak, regarded as representing the onset of permanent deformation, is called the yield point. For some materials, the stress–strain curve does not exhibit an obvious yield point but the curve does exhibit some feature which may be identified with the onset of permanent deformation, such as a rapid change of slope (see dashed line in Fig. 7.5). The stress at which this occurs may then be regarded as the yield stress. Plastics which fail at strains beyond yield are termed ductile. On stretching beyond yield, the plastic may neck and will eventually break at its ultimate tensile strength (which may be greater or less than the yield stress). Other plastics break at relatively low strains before reaching a yield point. These materials are termed brittle.

For materials which fail in a ductile manner, design for strength can be based on the tensile yield strength of the material. Thus, for multiaxial loading situations the tensile yield stress is used with a suitable yield criterion to predict component 'failure'.[2] If the material fails in a brittle manner during the tensile test, the stress at break cannot be used in the above approach.

Unfortunately, for thermoplastics, many factors can cause a nominally ductile material to behave in a brittle manner.[9] Thus, a reduction in temperature or an increase in the speed at which loads are applied to a specimen can promote a ductile-to-brittle transition. (Alternatively, an increase in temperature or a decrease in rate of loading can sometimes cause a brittle thermoplastic to fail by yielding.) Physical features such as sharp internal corners, voids, and surface defects can promote brittle failure in components made from nominally ductile materials. Other factors which can promote a ductile-to-brittle transition in a thermoplastic include hostile environments (e.g. chemical attack, outdoor exposure) and the use of inappropriate processing conditions during component manufacture.

The subject of brittle failure and the prediction of premature brittle failure in nominally ductile thermoplastics are currently receiving much attention. The fracture mechanics approach has been shown to be of value.[1-3, 10, 11] However, the application of fracture mechanics to short fibre reinforced thermoplastics is in its infancy and so will not be pursued further in this chapter.

It is possible for a single material to exhibit the complete range of curve shapes given in Fig. 7.5, when tests are carried out over a range of temperatures and/or extension rates. Thus both the 'failure' stress and the mode of failure vary with the conditions of test. Strength data derived from constant extension rate tests are therefore of limited applicability in component design.

For design purposes, it is desirable that the strength data are derived from tests in which the loading history matches as closely as possible that which is expected in service. Only the case of static loading will be considered here.

The strain response of a thermoplastic to the application of a constant force is described in Section 7.2.1. If the force is of sufficient magnitude, the specimen will eventually fracture. The time to fracture increases as the applied force decreases. This behaviour is known as 'creep rupture'. The data are usually presented as a graph of applied stress versus log of the time to fracture.[12] Fracture may be preceded by one or more of a number of events, each of which may be regarded as representing 'failure' according to the application. These events include crazing, stress whitening and local inhomogeneous deformation (deformation bands, necking). (These various 'failure' modes may also be observed for other loading histories.) Stress-log (time) graphs for each event provide valuable additional information on the material.

Data from creep rupture experiments are often superposed on the isometric stress–log (time) cross-plots of creep data. Some indication of the magnitude of the strains involved for each failure process may then be obtained. A typical presentation is shown in Fig. 7.6. The failure mode occurring should be noted on each curve.

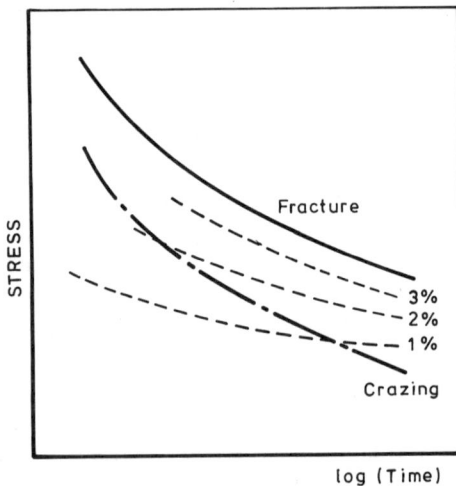

Fig. 7.6. Data from creep rupture experiments superposed on isometric cross-plots from creep data.

The creep rupture behaviour of a thermoplastic will depend on temperature and all the other factors discussed in relation to constant extension rate testing. The failure mode (ductile or brittle) may also change with time under load for any one set of test conditions. If ductile behaviour can be expected, then design for strength could follow the procedure mentioned under constant extension rate. The appropriate creep rupture data would be a plot of applied stress versus time to the onset of creep yield. The time under load in service would then be used to extract a value of yield stress from the data. Other ways of using the data are discussed elsewhere.[3]

## 7.3. ADDITIONAL CONSIDERATIONS FOR SHORT FIBRE REINFORCED THERMOPLASTICS

### 7.3.1. Fibre Volume Fraction and Fibre Length

The incorporation of short stiff fibres into thermoplastics introduces two additional basic parameters which affect the creep and creep rupture behaviour (and other mechanical properties). These parameters are the fibre volume fraction and the fibre length (or, from a fundamental viewpoint, the fibre aspect ratio).

The fibre volume fraction is fixed by the selection of the required material grade and this parameter does not, therefore, introduce any uncertainty into the selection of appropriate design data.

Fibres are often incorporated into thermoplastics by extrusion compounding.[13-16] This leads to a distribution of fibre lengths in the compounded material. Some further fibre breakage may then occur during injection moulding of the composite.[15,17] In general, the majority of the fibres in moulded components are in the length range from 0·05 mm to 1·0 mm and the individual fibres are fairly well dispersed.[18] Mean (number-average) fibre aspect ratios are normally in the range from 10:1 to 45:1, depending on base polymer and production route. Unfortunately, this is precisely the range in which the composite mechanical properties depend markedly on fibre aspect ratio. The possible influence of the injection moulding process on the fibre length distribution therefore introduces some uncertainty into data selection at the design stage. To a first approximation, this influence of processing on fibre length, and hence on the creep behaviour, of injection moulded components is taken into account by using specimens produced by injection moulding of the compound granules when generating the creep design data.

Alternative procedures are sometimes used to produce compound 'granules' in which the fibres are all of approximately the same length and equal to the 'length' of the granule.[13, 15] Fibre breakage will then occur during injection moulding to an extent which will be rather more dependent on the moulding conditions employed.[19] The test specimens for use in creep data generation will often be injection moulded under conditions designed to minimise fibre breakage in order to derive maximum benefit from the fibres. The use of such data in component design then assumes that similar care is taken during component manufacture.

### 7.3.2. Fibre Orientation

A far greater problem in data presentation and design with short fibre reinforced thermoplastics arises from the behaviour of the fibres during component fabrication. For example, during injection moulding, the flow of the fibre filled melt into the mould results in a complex fibre orientation distribution which cannot be significantly altered by any subsequent 'annealing' treatment. The variation of the fibre orientation distribution both through the moulding thickness and from place to place in the moulding leads to both anisotropy and inhomogeneity of mechanical properties.[20-24] In thick section mouldings, extensive voiding in the central core layer may add to the problem.[25]

The variation of fibre orientation through the thickness of a simple 2 mm thick, corner-gated square plaque injection moulded in glass fibre filled polypropylene is shown in Fig. 7.7. The photograph was produced by the application of the technique of X-ray contact microradiography[26] to a thin slice cut from the plaque at the position shown in Fig. 7.8. To a first approximation, the fibre orientation pattern can be divided into three layers, the plane of each layer being parallel to the plane of the plaque moulding. The two outer (or 'skin') layers contain fibres whose orientation is approximately random in the plane of the plaque. The central 'core' layer consists of fibres which are also reasonably well aligned in the plane of the moulding but, in this plane, the fibres have a preferred orientation which is *transverse* to the main direction of mould filling. This transversely aligned core layer occupies about 15% of the plaque thickness.

The 100 s tensile creep moduli of specimens cut from the plaque at $0°$ and $90°$ to the main direction of mould filling (see Fig. 7.8) are presented in Table 7.1. It is apparent that the tensile stiffness transverse to the mould fill direction, $E(90°)$, is marginally higher than that parallel to

TABLE 7.1
The 100 s Tensile Creep Modulus, $E$, of
Specimens Cut from Corner-Gated Square
Plaques, Measured at Low Strain at 23 °C[21]

| Material[a] | Plaque thickness (mm) | $E(GPa)$[b] (0°) | (90°) |
|---|---|---|---|
| GFPP | 2 | 2·80 | 3·20 |
| GFPP | 6 | 2·65 | 3·75 |
| GFPA | 2 | 5·55 | 3·65 |
| GFPA | 6 | 5·3 | 6·25 |

[a]GFPP, short glass fibre reinforced polypropylene; GFPA, short glass fibre
reinforced nylon 66.
[b]The 0° and 90° directions are defined in
Fig. 7.8.

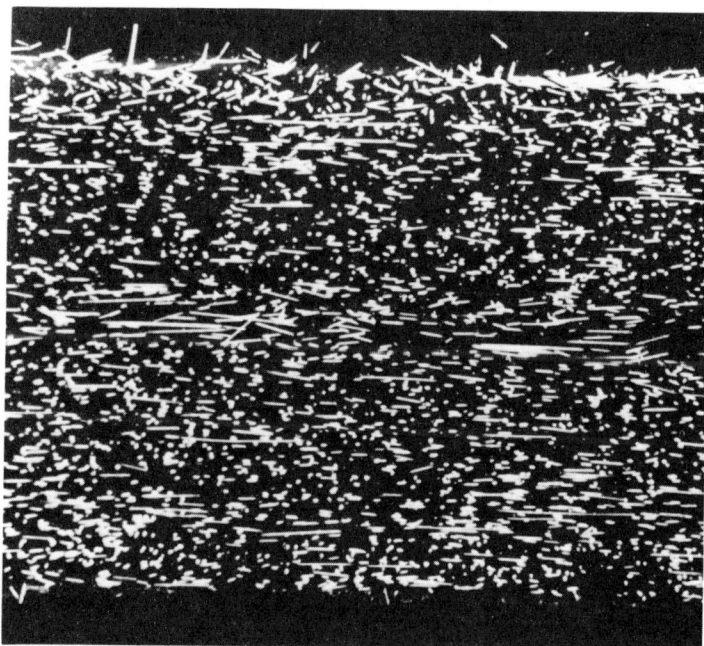

**Fig. 7.7.** Contact microradiograph showing fibre orientation through the
thickness direction for a 2 mm thick corner-gated square plaque injection
moulded in GFPP. (See Fig. 7.8 for site of slice for CMR examination.)

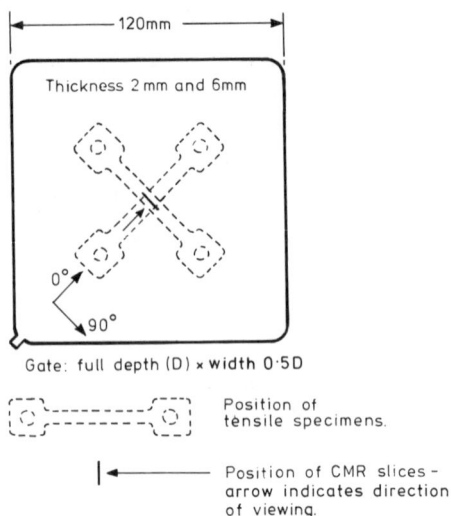

Fig. 7.8. Corner-gated square plaque. Definitions of 0° and 90° directions; sites of tensile test specimens and slice for contact microradiographic (CMR) analysis.

the mould fill direction, $E(0°)$. This is a direct consequence of the fibre pattern described above which shows a preferred orientation transverse to the main flow direction when the fibre orientation distributions in the various layers through the thickness are summed. (It should be noted that the fibre orientation distribution cannot be summed in this simple way when *flexural* modulus is being considered.[27])

If the plaque thickness is increased to 6 mm, a preferred orientation transverse to the main flow direction is obtained over most of the moulding thickness. The central core of highly transversely aligned fibres is then more difficult to distinguish but appears to occupy some 25–30% of the thickness. This increased preference for transverse alignment (compared with the 2 mm plaque) results in significant mechanical anisotropy for the 6 mm glass fibre filled polypropylene plaque, with $E(90°)$ being very much greater than $E(0°)$, as shown in Table 7.1.

The fibre orientation pattern observed in a given mould geometry depends on the base polymer. Thus, if the 2 mm thick corner-gated plaque is moulded in glass fibre filled nylon 66, the pattern shown schematically in Fig. 7.9 is obtained. In this instance, the core of fibres

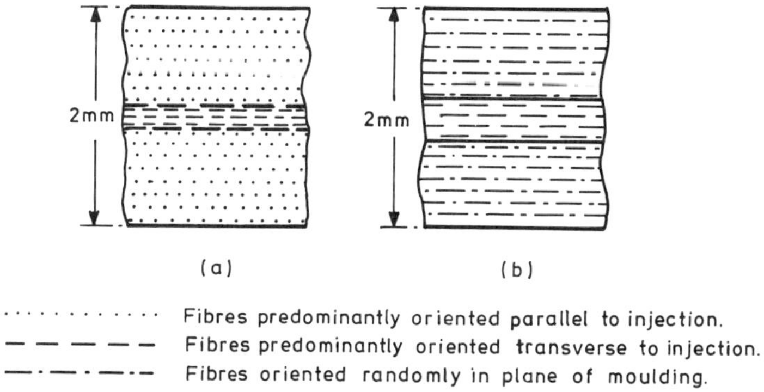

(a)                                   (b)

··········· Fibres predominantly oriented parallel to injection.
— — — — — Fibres predominantly oriented transverse to injection.
—·—·—·— Fibres oriented randomly in plane of moulding.

**Fig. 7.9.** Schematic representation of fibre orientation through the thickness direction of the 2 mm thick corner-gated plaque moulded in (a) GFPA and (b) GFPP (derived from Fig. 7.7 for comparison purposes).

aligned transverse to the main mould filling direction occupies only about 10% of the plaque thickness and the outer layers are more highly aligned in the $0°$ direction. Summing the fibre orientation distribution through the moulding thickness should, therefore, give an overall preferred orientation of the fibres in the $0°$ direction. This is confirmed by the modulus data included in Table 7.1, which show $E(0°) > E(90°)$ for the 2 mm thick glass filled nylon plaque.

When the 6 mm thick plaque was moulded in glass filled nylon 66, the outer (skin) layers still showed a preferred alignment of the fibres in the $0°$ direction but the core of highly transversely aligned fibres had increased to occupy about 50% of the moulding thickness.

The modulus data of Table 7.1 show that, for the 6 mm glass filled nylon plaque, $E(90°)$ is greater than $E(0°)$, i.e. the overall through-thickness fibre orientation distribution is biased towards alignment of the fibres transverse to the main direction of mould filling. This is in marked contrast to the behaviour of the 2 mm plaque.

For each base polymer, mould geometry has a significant influence on the fibre orientation distribution. The effect of changing mould thickness, described above, is one example of this influence. The patterns of fibre orientation described above are typical of those generally observed in diverging flow situations, i.e. those in which the fibre filled melt experiences a significant lateral expansion as it advances. If the melt experiences a lateral contraction as it advances (convergent flow), the fibres show a strong tendency to align parallel to the main flow direction. However, a core of transversely aligned fibres is a common feature in injection moulded parts.[22]

Moulding conditions can influence the fibre orientation distribution in a given mould geometry but, in general, this influence is of less significance than that of mould geometry itself.[21]

The fibre orientation distributions in the above mouldings have been approximated to a three-layer structure through the moulding thickness. In some instances, an additional two layers were discernible (i.e. one on each surface).[21] More complex layered structures have also been reported.[20, 24]

## 7.4. THE PRESENTATION AND SELECTION OF STIFFNESS AND STRENGTH DATA FOR USE IN DESIGN

The additional considerations discussed in Section 7.3 complicate even further the complex situation outlined in Section 7.2 and emphasise the need for new, simplified approaches to data generation and presentation for short fibre reinforced thermoplastics.

### 7.4.1. The Bounds Approach to Data Presentation

In general, tensile stiffness and tensile strength data have been produced using injection moulded, dumb-bell shaped tensile specimens. In the case of unfilled thermoplastics (or homogeneous, isotropic filled thermoplastics) data derived from tests on these specimens will be reasonably representative of the properties of the material when moulded into components of more complex geometry. (This assumes that due consideration is given to factors such as molecular orientation and processing/storage history.)

The production of standard tensile test bars by the injection moulding of short fibre reinforced thermoplastics gives test specimens in which the fibres are aligned predominantly along the tensile axis.[18] Data from tests on these bars therefore approach the best (i.e. highest tensile stiffness and strength) that can be expected of the material. For this reason, such data are often referred to as the *upper bound* of behaviour.[23] Even if such alignment occurs in a moulded article, it is quite possible that the major stress component will be in a direction transverse to the direction of predominant fibre alignment. The tensile stiffness and strength in this transverse direction will be much lower than that obtained in tensile tests carried out parallel to the direction of fibre alignment. Tensile tests carried out at intermediate angles would normally give values lying between these two cases.

It is apparent that tensile tests carried out on specimens cut with their tensile axis at 90° to the axis of standard tensile bars would give stiffness and strength values approaching the lowest that could normally be expected of the material. Such data are often referred to as the *lower bound* of behaviour.[23] It is evident that such tests are not normally possible and that an alternative source of lower bound specimens is required. For thermoplastic polyesters and polyamides, a rectangular 100 mm × 100 mm × 2 mm plaque, film-gated along one edge, has been found to be useful for this purpose.[21, 28] Thus, by careful selection of moulding conditions, it is possible to produce a moulding in which there is a reasonable degree of fibre alignment in the major flow direction. Tensile test specimens cut perpendicular (90°) to the flow direction can then be used to produce data that give a reasonable representation of the lower bound behaviour of the material. The plaque moulding is not suitable for use with polypropylene[21] and alternative mouldings have been investigated.[23, 28]

Tensile test programmes covering the range of variables outlined in Section 7.2.3 can be carried out on both the upper and lower bound specimens. This provides an indication of the range of behaviour likely to be encountered in moulded articles but can involve considerable experimental effort, especially if data on grades covering a range of fibre volume fractions are needed. For stiffness, procedures have been developed for the calculation of composite creep behaviour based on a knowledge of the creep behaviour of the matrix and fibre properties.[28] Careful application of these predictive tools and interpolation procedures can result in useful savings in experimental effort.

### 7.4.2. The Choice of Appropriate Data for Design Purposes

The difference in behaviour between the upper and lower bounds of tensile modulus and tensile strength can be relatively large and some guidance must be given to the designer on the selection of appropriate modulus and strength values. (In contrast, the in-plane shear modulus shows far less variation with direction of test, even for mouldings in which there is a high degree of fibre alignment.[29])

With the very limited information on fibre orientation available to them at the time, Dunn and Turner[23] suggested that the mean of the bounds of tensile modulus might provide a useful compromise in deformation design.

Recent research has greatly increased our knowledge and understanding of the way in which the fibre orientation in a moulding is influenced by the mould geometry,[15, 21] processing conditions and the base polymer.[30] However, at the present time, the detailed fibre orientation distribution in a moulded part can still not be predicted with any certainty at the design stage. Even if this were possible, full allowance in the design calculations for the changing pattern of anisotropy over the component could not normally be justified on economic grounds.

The results of fibre orientation studies on a range of components have shown that a strong preferred fibre alignment in one direction throughout a component is unlikely.[22] Rather, fibre alignment in different directions in layers through the thickness will often result in only a weak overall preferred alignment in many parts of the moulding. Furthermore, regions of high fibre alignment will often be balanced by other regions in which the preferred fibre alignment direction is markedly different. The general deformation behaviour of a component will be controlled by some average of the response of the various regions of differing orientation (and hence, differing moduli) over the area of interest. It was, therefore, proposed that in many instances it would be realistic to use creep data corresponding to a fibre distribution in which the fibres were oriented randomly in the plane of the moulding and to carry out design calculations assuming the material to be isotropic.[21, 31]

Random-in-the-plane modulus data can be readily calculated from the bounds data or predicted from a knowledge of the creep behaviour of the base polymer.[28]

From a consideration of orthotropic materials, Stephenson et al.[32] suggested that a 'mean stiffness' was useful for materials characterisation and in deformation design. Modulus values derived using this approach will not differ greatly from those corresponding to a random-in-the-plane fibre orientation distribution.

It is inevitable that the use of a random-in-the-plane modulus will introduce some error into the design calculations in many instances. However, when seeking to reduce this error it must be remembered that there are many other factors which may contribute to a discrepancy between design prediction and actual component behaviour. These include the influence of processing and storage history on modulus and the difficulty of predicting the effects of complex stress and temperature histories which occur in service.

The strength behaviour of a moulded component will not usually be related to an average of the fibre distribution. Failure may be initiated at a point in the moulding where, for example, the fibres happen to be highly aligned in a direction transverse to a major stress component. This suggests that, in critical areas, the possibility of failure by, for example, creep rupture should be checked using creep rupture data for the lower bound, unless the fibre distribution can be assumed with certainty to be more favourable.

Alternative proposals for the generation of stiffness and strength design data, based on the use of 'sub-components' and de-rating factors, have been advanced.[33, 34] Unfortunately, no examples of the use of such data in component design are available.

## 7.5. DESIGN FOR STIFFNESS WITH SHORT FIBRE REINFORCED THERMOPLASTICS

The prediction of the deformation of components moulded in short fibre reinforced thermoplastics does not differ in principle from the prediction of the performance of articles made from other materials. The procedure involves identifying a suitable structural idealisation for the product and from this establishing a relationship between deformation and applied load. This relationship between deformation and applied load can be used, along with a stiffness value appropriate to the operating conditions, to predict deformation for a given load. The difference in emphasis for short fibre reinforced thermoplastics centres on the increased difficulty of selection of appropriate stiffness data.

In the previous section, the use of creep data corresponding to a 'random-in-the-plane' fibre orientation distribution in conjunction with standard 'isotropic' analysis was discussed. This was considered to represent a useful compromise between accuracy and ease of use in many design situations. It is apparent that confidence in any design procedure can only result from experience of its application. Unfortunately, there is still a lack of published work on short fibre reinforced thermoplastics which permits a critical evaluation of stiffness design procedures for injection moulded components. Some results from a programme carried out at Cranfield Institute of Technology are, therefore, described below.

The Cranfield programme was directed towards an examination of the question of data selection for the design of injection moulded short fibre reinforced thermoplastic components. The approach adopted was to take injection moulded components, or subcomponents, and to subject them to various loading configurations. Experimental load versus deflection results from these components were compared with the behaviour predicted from the relevant stress analysis for the particular geometry and loading combinations.

In all cases, results were predicted using tensile modulus data appropriate to the following three situations.

(A) Fibres highly aligned uniaxially and stress applied in the direction of fibre alignment (upper bound approach).

(B) As (A), but stress applied in a direction at right angles to the fibre alignment direction (lower bound approach).

(C) Fibres oriented randomly in the plane of the moulding (random-in-the-plane, or RITP, approach).

The efficiency of any stress analysis is dominated by the accuracy of the initial input data. Stiffness data for predicting component behaviour were therefore produced at Cranfield using highly acccurate creep machines[35] and following the procedures recommended in British Standard BS4618.[4] The moulding and storage conditions of the test specimens used for producing design data were matched as closely as possible to those of the components being studied.

Whenever possible, the component under study was also moulded in an unreinforced grade of the material and the experimental load–deflection behaviour compared with prediction. As considerable care was taken to try to ensure that the polymer in the specimens used for design data production was in the same 'molecular state' as the moulded components, the major uncertainty for the unreinforced component was concerned with the stress analysis and not the materials data. The tests on the unreinforced component therefore act as a check on the accuracy of the stress analysis used.

Two examples from this study are presented below.

### 7.5.1. Example A: Subcomponent from a Large Container Moulding

The container shown in Fig. 7.10 was injection moulded in unreinforced polypropylene and polypropylene containing 30 wt % of short glass fibres (GFPP). The container has been used as a source of a number of subcomponents suitable for the study outlined above. The region of the side wall between the top and bottom flanges and the two vertical ribs has been used as one such subcomponent.

The component was clamped on a purpose-built frame designed to provide a flat rectangular plate with its edges fixed to the frame by the ribs and flanges, as shown in Fig. 7.11. The central deflection of the plate, produced by a uniformly distributed load, was measured by means of highly accurate capacitance transducers and the experimental load–deflection characteristics produced. The load was provided by means of very fine lead shot, held within a thin rubber membrane for ease of loading.

Isochronous 100 s test procedures[4] were used to produce the experimental load–deflection results for the components moulded in both the

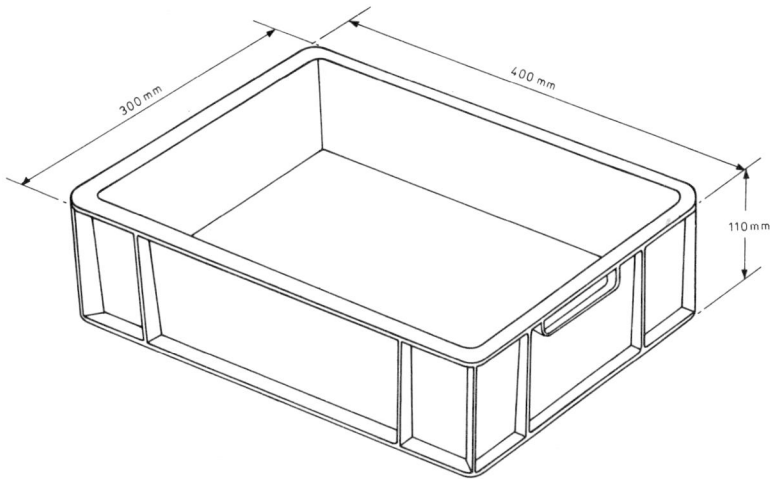

Fig. 7.10. Large container moulding used in deformation studies.

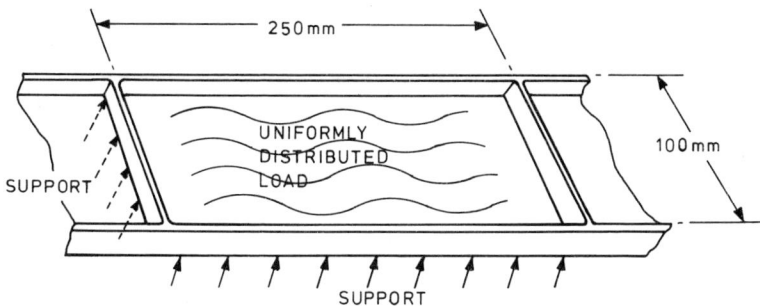

Fig. 7.11. Subcomponent taken from the large container moulding of Fig. 7.10.

filled and unfilled materials. All experimental results were obtained at a test temperature of 23 °C. In all cases, deflections below one half the thickness of the plate were studied in order that 'small' deflection stress analysis equations could be used to predict the component behaviour.

The load–deflection behaviour was predicted from standard elastic flat plate equations of the form:

$$\delta = \frac{\alpha W b^3}{a d^3 E}$$

where $\delta$ = central deflection of the plate,
$W$ = total applied force, uniformly distributed,
$b$ = plate width,
$a$ = plate length,
$d$ = plate thickness,
$E$ = tensile modulus,
$\alpha$ = fixity (constant).

The constant $\alpha$ is dependent upon the type of edge restraints and the dimensions of the plate. Values of $\alpha$ are generally quoted for the two extreme cases of the edges being solidly clamped (i.e. deflection at the edge is zero), or for the edges being merely supported from underneath (i.e. the corners are allowed to lift under the applied load). In practice, an edge restraint somewhere between these two extremes is achieved.

The results for the unreinforced polypropylene component were used to overcome the uncertainties in edge restraint and to allow for any additional errors due to the stress analysis. Thus, the experimental force–deflection results (shown in Fig. 7.12), together with the measured 100 s tensile creep modulus for the unfilled polypropylene, were used to obtain a value of the fixity, $\alpha$, applicable to the test apparatus used. This procedure gave a fixity value of 0·042 which, as expected, lies between the two theoretical extremes of 0·0277 and 0·1110 quoted in Ref. 36. It has been assumed that the value of $\alpha$ calculated for the unreinforced polypropylene component will also apply to the glass fibre reinforced polypropylene component.

The force–deflection behaviour of the glass fibre reinforced polypropylene component has been predicted using experimental values of 100 s tensile creep modulus appropriate to the three situations, (A), (B) and (C), defined above. The three predicted responses are compared with the experimental force–deflection response of the glass fibre reinforced polypropylene component in Fig. 7.13. It is apparent that at

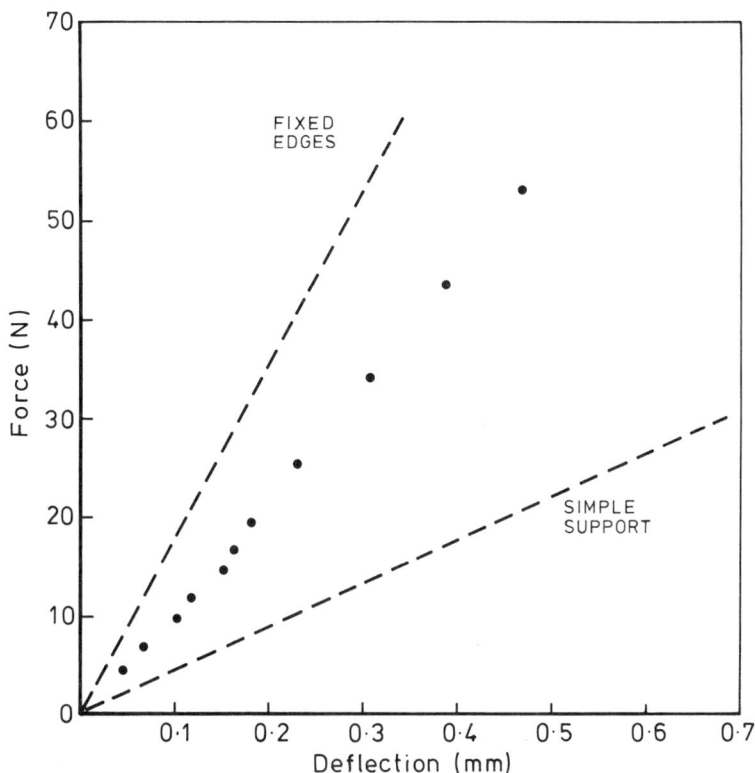

Fig. 7.12. ●, Experimental force–deflection behaviour of the container subcomponent moulded in polypropylene (PP). The dashed lines show the behaviour predicted assuming that the edges of the component are either rigidly fixed or simply supported.

very low deflections (less than 0·15 mm), the prediction assuming a modulus appropriate to a random-in-the-plane fibre orientation distribution is in good agreement with the experimental results. At higher deflections, the experimental results deviate from the RITP prediction towards the lower bound prediction. At all deflections, the upper bound prediction (based on tensile bar data) gives a considerable underestimate of deflection.

It should be noted that modulus data applicable to a 100 s strain of 0·1% were used in the above calculations. The maximum strain in the component at the highest deflection presented in Fig. 7.13 ($\delta \sim 0.28$ mm)

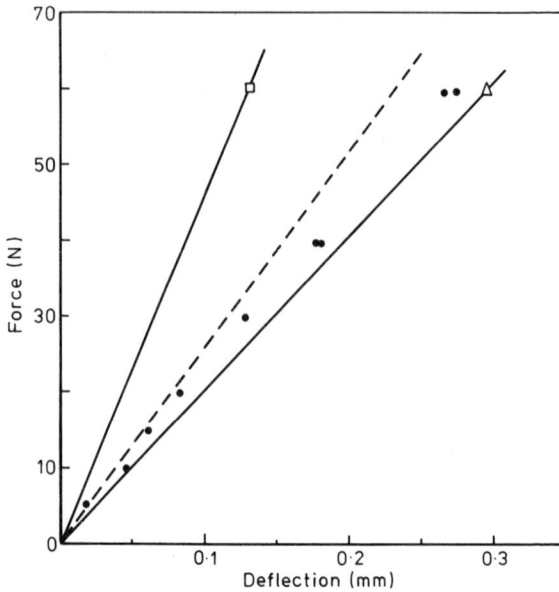

**Fig. 7.13.** Experimental and predicted force–deflection behaviour of the container subcomponent moulded in glass fibre reinforced polypropylene (GFPP). ●, Experimental data. Predictions based on use of data for the upper bound (□), lower bound (△) and random-in-the-plane fibre orientation distribution (---).

was 0·12%. As the behaviour of the material in this strain region approximates to linear viscoelastic, the deviation between the RITP prediction and experiment cannot be attributed to non-linear viscoelastic effects in this instance.

### 7.5.2. Example B: Gear Selector

The injection moulded 'gear selector' shown in Fig. 7.14 provided the component for the second study.[37] This component had already been the subject of a design study by Courtaulds Carbon Fibre Research Division, and an injection moulding tool had been manufactured for the purpose of that study. This tool was loaned to Cranfield for the duration of the work to enable components to be produced in a number of materials. The availability of the tool also enabled the processing conditions and storage histories of all components to be controlled as required.

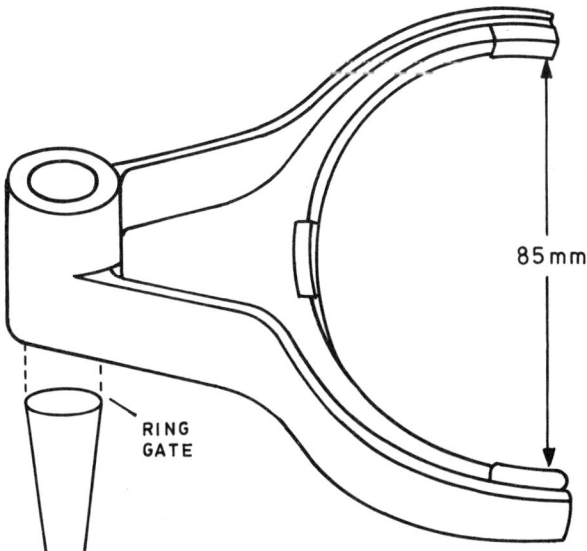

**Fig. 7.14.** Schematic representation of the gear selector component. (From Ref. 37.)

The injection moulding tool was single cavity incorporating a sprue feed through a ring gate as shown in Fig. 7.14. The gear selector was moulded in a number of different materials. However, for the purpose of this example, only the results of tests carried out on components moulded in polybutylene terephthalate (PBT) and PBT reinforced with 30 wt % of short glass fibres (GFPBT) are presented.

The gear selectors were produced at Cranfield Institute of Technology on a Bipel 130/25 injection moulding machine rated at 130 ton clamp force and 180 ml shot volume. This was fitted with a Bosch SPR 200 closed-loop control system. All components were moulded under standard processing conditions recommended for the various materials.

Specially designed fixtures were built to allow the gear selectors to be loaded in the directions $X$ and $Y$ shown in Fig. 7.15. Isochronous 100 s tests were carried out on the components, moulded in both the filled and unfilled grades of material, at a test temperature of 23 °C.

Capacitance transducers were used to measure deflections $dX$ and $dY$ corresponding to the two loading directions as shown in Fig. 7.15.

A theoretical analysis of the gear selector was carried out by means of a finite element (FE) stress analysis. ANSYS, a commercially available

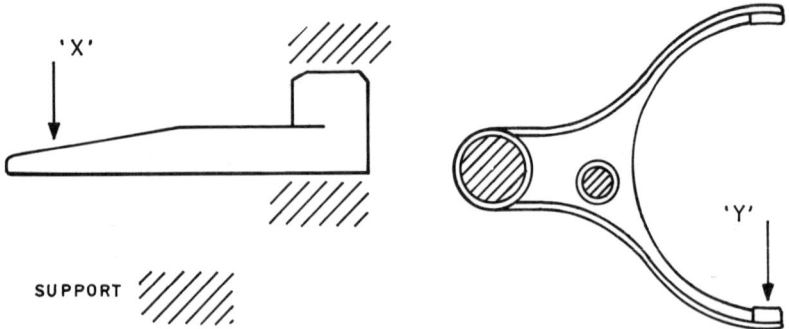

**Fig. 7.15.** Loading modes $X$ and $Y$ for the gear selector. (From Ref. 37.)

FE package, was used and a linear analysis carried out using eight-noded linear–parabolic elements to model the component. Symmetry about the component centreline was assumed and where it was impracticable to allow for small radii, such as at the very tip of the forks, a square profile was assumed. These assumptions are standard practice with FE programmes and should not detract from the accuracy of the analysis.

For components loaded in direction $X$ a unit load was assumed, acting as a point load, 7·5 mm from the tip of the forks. Loading in direction $Y$ was applied as a unit pressure, distributed over one element at the tip of the fork.

The FE analysis was performed using a tensile modulus, $E$, of 13 GPa, for the case of the load applied in direction $X$, and tensile moduli of 13 GPa and 3 GPa for loading in direction $Y$. In the case of direction $Y$, two moduli were used as a check on the linearity of the results. The relationship

$$\delta E = \text{constant}$$

was found to apply for the two modulus values ($\delta$ being the deflection at any chosen nodal point). This being the case, it was possible to use the relationship corresponding to the appropriate nodal point to predict deflection for any value of tensile modulus.

Stiffness data for the glass reinforced PBT were obtained from ISO II dumb-bells (upper bound data) and dumb-bell specimens cut at 90° to the melt flow direction from 2 mm thick flash-gated plaques (lower bound), as shown in Fig. 7.16. Modulus data for a random-in-the-plane

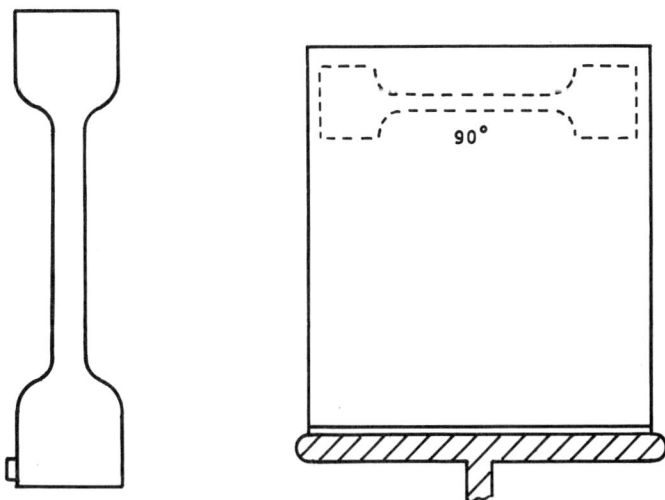

**Fig. 7.16.** ISO II dumb-bell and flash-gated plaque mouldings used for production of creep data on glass fibre reinforced polybutylene terephthalate (GFPBT). (From Ref. 37.)

fibre orientation distribution were calculated from the upper and lower bound data. Specimens cut from flash-gated plaques at 90° to the flow direction were used to produce data for the unfilled PBT. Low-strain 100 s tensile modulus data at 23 °C are shown in Table 7.2.

The finite element analysis was used to predict the force–deflection behaviour of the gear selector, in loading modes $X$ and $Y$, using the modulus values appropriate to both the unfilled and glass fibre

TABLE 7.2
The 100 s Tensile Creep Modulus for PBT and
GFPBT at 23 °C (100 s Strain ~0·1%)

| Material | Specimen | Modulus (GPa) |
|----------|----------|---------------|
| PBT | Plaque (90°) | 2·5 |
| GFPBT | ISO II bar | 10·0 |
| GFPBT | Plaque (90°) | 5·0 |
| GFPBT | RITP[a] | 7·0 |

[a]Value calculated for a random-in-the-plane fibre orientation distribution.

reinforced PBT. The prediction for the unfilled PBT was carried out in order to examine the accuracy of the analysis. For the glass fibre reinforced PBT, modulus data corresponding to the three situations (A), (B) and (C) defined above were used in the calculations.

Predicted and experimental force versus deflection results for the gear selector loaded in direction $X$ and subjected to the 100 s isochronous test procedure are presented in Fig. 7.17. The comparison between experimental and predicted results for the unfilled PBT component demonstrates the accuracy of the finite element stress analysis. It is evident from the results that close agreement between prediction and experiment is obtained with the components moulded in the unfilled material. The results for the glass filled material clearly show that if upper bound modulus data (as obtained from standard tensile test bars) are used in the FE predictions, considerable underestimation of deflection for a given load will result. If modulus data appropriate to the lower bound is used, then an overestimate of deflection is obtained. However, it is apparent that the experimental results lie closer to the lower bound predictions than to the upper bound prediction. The predicted behaviour assuming a modulus appropriate to a random-in-the-plane fibre orientation distribution gives close agreement with experiment.

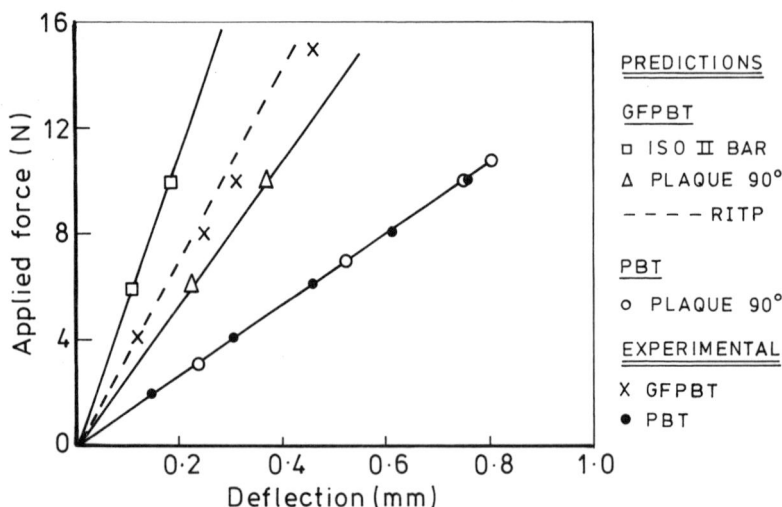

Fig. 7.17. Isochronous 100 s force–deflection behaviour at 23 °C for PBT and GFPBT gear selectors loaded in direction $X$. (From Ref. 37.)

A similar presentation to that discussed above, but for loading direction *Y*, can be seen in Fig. 7.18. Again, the results for the unfilled PBT confirm the accuracy of the stress analysis. For the glass fibre reinforced PBT, the comparison between experiment and prediction follows the pattern found for loading direction *X*, the RITP approach giving the closest agreement with experiment. The small deviation between the RITP prediction and experiment as deflection increases mirrors the trend apparent in the predicted and experimental data for the unfilled PBT.

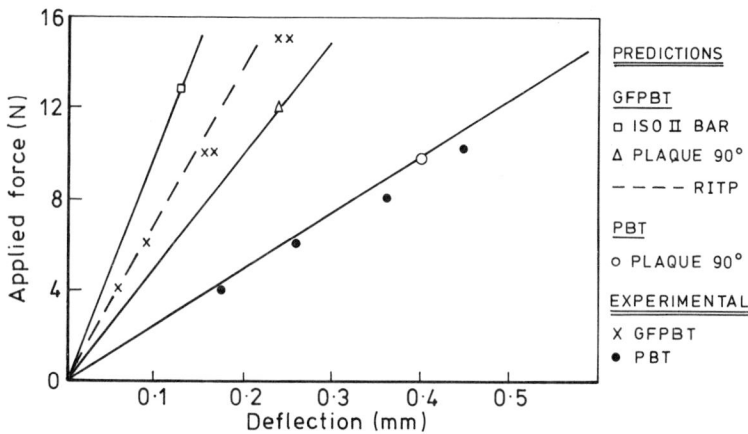

**Fig. 7.18.** Isochronous 100 s force–deflection behaviour at 23 °C for PBT and GFPBT gear selectors loaded in direction *Y*. (From Ref. 37.)

### 7.5.3. Discussion

Significant differences in fibre orientation distribution between GFPP and GFPA, for a given mould geometry, were noted in Section 7.3.2. The fibre orientation patterns found in GFPBT mouldings are generally similar to those for short glass fibre reinforced nylon 66 (and nylon 6). Significant differences may therefore occur in the fibre orientation distributions for GFPP and GFPBT in some mould geometries. It is therefore important to consider mouldings in both types of material when evaluating design procedures.

The study on the gear selector described in the previous section has been repeated for gear selectors moulded in polypropylene and polypropylene reinforced with short glass fibres.[37] The general pattern of the results was similar to that described for the PBT and GFPBT gear selectors.

The GFPP container moulding described in Section 7.3.1 has been the subject of a number of studies. As well as tests on the whole component, several different subcomponents were extracted from the container and subjected to a variety of loading modes. Short studies have also been carried out on several other components moulded in GFPBT and GFPP.

In all the above studies, the deformation response predicted using Young's modulus data appropriate to a random-in-the-plane fibre orientation distribution (and assuming the material to be homogeneous and isotropic) gave reasonable agreement with experiment.

## 7.6. DESIGN FOR STRENGTH

The random-in-the-plane fibre orientation distribution approach to design with short fibre reinforced thermoplastics gives the designer a simple and effective method for low-strain deformation design. However, the problem is not as simple where design to failure is concerned. It has, in the past, been suggested that lower bound data should be used along with the appropriate formulae to predict failure loads.[21, 31] If failure under load were controlled purely by fibre orientation distribution, this would seem a justifiable approach, combining ease of use with a degree of safety.

Since the above proposal was put forward, studies carried out on a number of simple components[37] have indicated that, in flexural situations the components could withstand considerably greater loads than those predicted using the relevant elastic formulae. The following two examples illustrate the errors associated with this procedure.

### 7.6.1. Example A: Injection Moulded, Ribbed Plaque

The ribbed plaque shown in Fig. 7.19 has been injection moulded in glass fibre reinforced polypropylene (GFPP) and glass fibre reinforced nylon 66 (GFPA). Sections cut from the plaque, as shown in Fig. 7.19, have been tested in a three-point bending situation (with rib in tension) under constant deflection rate loading. The test temperature was 23 °C. The maximum force that each section could withstand before complete failure occurred was measured experimentally. This has been compared with the predicted force to failure, calculated assuming that the

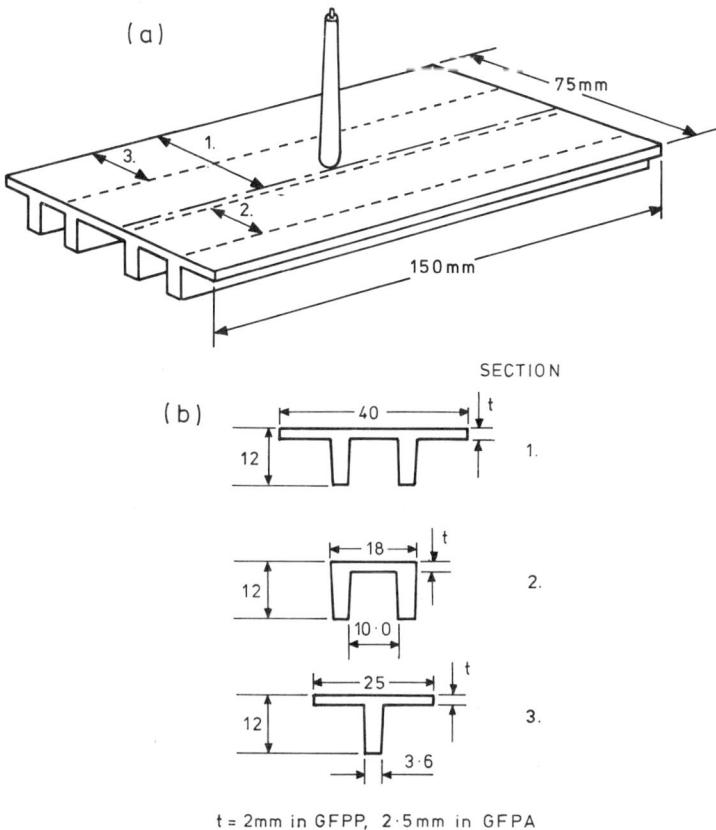

**Fig. 7.19.** Schematic representation of the ribbed plaque component showing (a) the sites of the three sections extracted from the moulding and (b) details of each of the three cross-sections.

component fails when the maximum stress within the component reaches the tensile strength of the material.

Tensile specimens, cut from the ribs of the plaque, have been used to produce the tensile strength data used in the predictions. The data for both GFPP and GFPA can be seen in Table 7.3.

A simple linear elastic approach has been used to predict the failure loads, eqn (7.5) being used in each calculation:

$$F = \frac{4I\sigma_u}{LC} \tag{7.5}$$

where $F$ = force at failure,

$\quad\quad I$ = second moment of area about the neutral axis,

$\quad\quad L$ = span,

$\quad\quad C$ = distance of outer (tensile) surface from neutral axis,

$\quad\quad \sigma_u$ = tensile strength of material.

The predicted and experimental results of force at failure are presented in Table 7.3. It is apparent from the results that the predicted forces are considerably lower than those which can be withstood in practice.

TABLE 7.3
Comparison of Predicted and Experimental Force at Failure for
Sections Taken from the Ribbed Plaques

Predictions based on a linear elastic analysis.

| Material | Rib section | Tensile strength[a] (MPa) | Force at failure (N) | |
|---|---|---|---|---|
| | | | Predicted | Experimental |
| GFPP | 1 | 50 | 375 | 681 |
| GFPP | 2 | 50 | 324 | 623 |
| GFPP | 3 | 50 | 201 | 375 |
| GFPA | 1 | 117 | 879 | 1 483 |
| GFPA | 2 | 117 | 759 | 1 383 |
| GFPA | 3 | 117 | 471 | 817 |

[a]Tensile strength of specimens taken from the plaque ribs.

It should be noted that the tensile strength data used above are well above the lower bound values for the materials. For example, specimens cut from the plaque base gave strengths of 32 and 79 MPa for GFPP and GFPA respectively. (These values correspond approximately to the values for a random-in-the-plane fibre orientation distribution.) Thus, if a true lower bound value was used in the prediction, even greater discrepancy between predicted and experimental results would occur. Even if upper bound data were used in the predictions, considerable underprediction of failure forces would still result. This suggests that the main problem lies not with the data selection but with the type of analysis used.

### 7.6.2. Example B: Gear Selector
The gear selector discussed in Section 7.5.2 has been used as a further example to examine the strength behaviour of components. As in

example A (Section 7.6.1), failure is deemed to have occurred when the maximum stress within the component reaches the tensile strength of the material. Gear selectors moulded in glass fibre reinforced polybutylene terephthalate (GFPBT) and glass fibre reinforced polypropylene (GFPP) were tested at constant extension rate on a tensile testing machine at 23 °C and the force at failure recorded. Only loading mode *Y* was used (see fig. 7.15).

The finite element stress analysis carried out on the component predicts that, at an applied force of 14·7 N, the maximum stress within the component is 8·14 MPa. On the assumption that the relationship between force and stress remains linear to failure, it is possible to predict the expected failure force corresponding to the tensile strength of the materials. A comparison between the predicted and experimental forces is given in Table 7.4. The predicted values have been calculated assuming tensile strength data appropriate to the upper bound and it is again apparent that the predictions are substantially lower than the experimental results. Clearly, if the lower bound data were used, the discrepancy between prediction and experiment would be even greater.

The same elastic analysis has also been used in conjunction with *flexural* strength data (as opposed to tensile data) to predict the failure force for the gear selector. The results, included in Table 7.4, indicate much closer agreement with experiment than was obtained when the tensile data were used. The predictions of Table 7.4 were made using upper bound flexural data and an overestimate of failure force has resulted for each material. However, this overestimate is only 8% for the GFPP and 12% for the GFPBT gear selectors.

TABLE 7.4
Comparison of Predicted and Experimental Force at Failure for the
Gear Selector Component

Predictions based on a linear elastic analysis.

| Material | Predicted force at failure (N) | | Experimental force at failure (N) |
|---|---|---|---|
| | Using TS data[a] | Using FS data[b] | |
| GFPP | 108 | 151 | 140 |
| GFPBT | 253 | 370 | 331 |

[a]Calculated using upper bound tensile strength data.
[b]Calculated using upper bound flexural strength data.

### 7.6.3. Comparison of Tensile and Flexural Strength

In order to understand further the mode of failure of components moulded in short fibre reinforced thermoplastics, an examination of the relationship between tensile and flexural strength data is required. Data obtained on injection moulded ASTM tensile dumb-bells and specimens at 0° and 90° to the main flow direction from 4 mm thick, corner-gated square plaques are presented in Table 7.5 for GFPBT. Test temperatures of 23°C and 60°C were employed. All testing was carried out on a constant extension rate test machine. The deflection rate used during the flexure tests was chosen such that the outer fibre strain rate corresponded to the nominal tensile strain rate used during the tensile tests.

TABLE 7.5
Tensile and Flexural Strength Data at 23°C and 60°C for GFPBT

| Test specimen | Temperature (°C) | Tensile strength (MPa) | Flexural strength (MPa) |
|---|---|---|---|
| ASTM bar | 23 | 138 | 205 |
| Plaque (0°) | 23 | 98 | 170 |
| Plaque (90°) | 23 | 95 | 124 |
| ASTM bar | 60 | 93 | 138 |
| Plaque (0°) | 60 | 66 | 114 |
| Plaque (90°) | 60 | 68 | 85 |

The results for the ASTM bars moulded in GFPBT (Table 7.5) indicate that the flexural strength is considerably higher than the tensile strength, for both test temperatures. This could be a consequence of the inhomogeneous fibre orientation distribution through the bar thickness or the occurrence of some degree of yielding (or 'plastic collapse') during the flexure tests (or a combination of both effects). It is not possible, from the bar data alone, to distinguish between the two possible causes. (Ideally, this requires that the tensile and flexural tests are carried out on specimens in which the fibre orientation distribution is homogeneous).

The results for the plaque specimens (Table 7.5) show that the flexural strength is greater than the tensile strength in *both* the 0° and 90° directions. The flexural strength is also anisotropic. As the flexural strength is dominated by the fibre orientation in the surface layers, the higher strength of the 0° specimens indicates a preferred alignment of the fibres in the $0°$ direction in the surface layers of the plaque. It is apparent from Table 7.5 that if the fibres had been oriented randomly in the plane of the moulding, a flexural strength in the region of 140 MPa at 23 °C would have been obtained in both the $0°$ and $90°$ directions in the plaque.

The tensile strengths of the plaque $0°$ and $90°$ specimens (Table 7.5) are very similar. To a first approximation, these tensile results reflect a through-thickness fibre orientation distribution in which the preferred fibre alignment in the $0°$ direction in the surface layers is balanced by a preferred alignment in the $90°$ direction in the central (core) layer. The results suggest that a homogeneous, random-in-the-plane fibre orientation distribution would give a tensile strength in the plane of approximately 96 MPa at 23 °C. This is considerably lower than the value of 140 MPa suggested above for the flexural strength for the same fibre orientation distribution.

The plaque results indicate that some factor other than fibre orientation must be providing the major contribution to the superior flexural strength values. This could be the occurrence of some degree of 'plastic collapse' during the flexure test. Components which are deformed in flexure in service, may similarly undergo some degree of plastic collapse prior to final 'failure'. In such cases, it would appear to be more appropriate to use flexural strength data in the design calculations. However, this approach may lead to errors when the thickness of the test specimens differs from the wall thickness of the component. One alternative is to use an elastic–plastic design approach.

### 7.6.4. Elastic–Plastic Analysis

The tensile stress-strain behaviour of an elastic-, perfectly plastic material is shown in Fig. 7.20. For such a material, the stress remains constant at the yield value, $\sigma_y$, as the strain is increased beyond the yield point.

As an example of elastic–plastic behaviour in a 'component', it is convenient to consider a rectangular beam subjected to pure bending (Fig. 7.21). The standard bending equation (eqn (7.3)) can be used until the maximum stress within the beam reaches $\sigma_y$ (Fig. 7.21(b)) where

$$\sigma_y = \frac{M_e C}{I}$$

$M_e$ is the maximum bending moment that can be maintained with the entire section of the beam still behaving in a totally 'elastic' manner. If

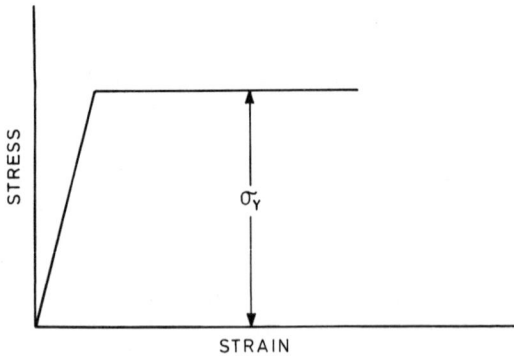

Fig. 7.20. Tensile stress–strain behaviour of an elastic, perfectly plastic material.

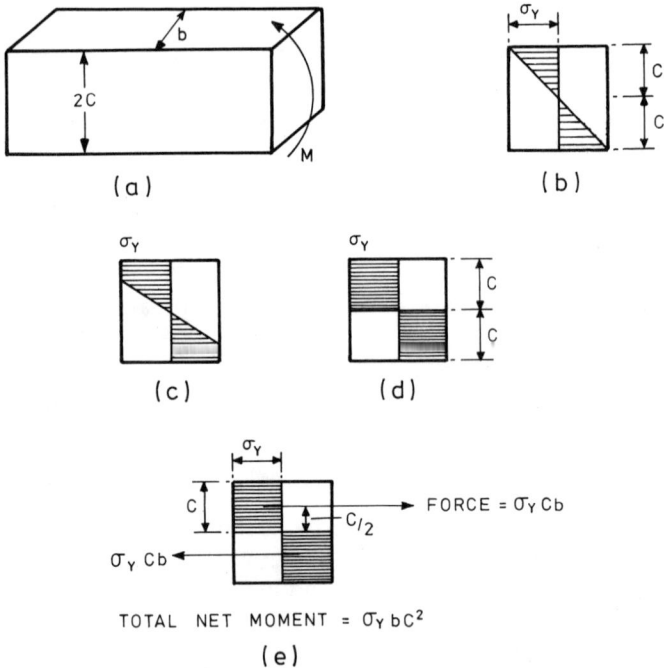

Fig. 7.21. Elastic–plastic behaviour of a rectangular beam subjected to pure bending. (a) Schematic showing beam loading and dimensions; (b) stress distribution at the limit of elastic behaviour; (c) partial plastic behaviour; (d) total plastic behaviour; (e) total moment at plastic collapse.

the load on the beam increases, producing a bending moment greater than $M_e$, then some degree of plastic behaviour starts to occur (Fig. 7.21(c)). As the load is increased further, the bending moment will increase until the total section is plastic (Fig. 7.21(d)). At this point, the section cannot withstand any further increase in load. The bending moment at this point is termed the limit moment or the moment of total plastic collapse. As the section has achieved a state of total plasticity, there must be a balance of forces acting on the section, along the beam axis. This being the case, the moment at the point of total plastic collapse, $M_p$, can be calculated as the total moment due to the net forces (Fig. 7.21(e)). In the case of a rectangular beam, this is given by

$$M_p = \sigma_y b C^2 \tag{7.6}$$

The 'total plastic collapse' approach has been applied to the specimens used in the tensile/flexural comparison of Section 7.6.3. For each type of specimen, the force at which total plasticity occurs in the three-point bend test has been calculated using the tensile strength value for that specimen type (Table 7.5). These predicted failure force values are compared with the experimental values of applied force at failure for the three-point bend tests in Table 7.6.

TABLE 7.6
Comparison of Predicted and Experimental Force at Failure for GFPBT Specimens Tested in Three-point Bend at Temperatures of 23 °C and 60 °C

Predictions based on a total plastic collapse analysis.

| Test specimen | Temperature (°C) | Force at failure (N) | |
|---|---|---|---|
| | | Predicted | Experimental |
| ASTM bar | 23 | 425 | 420 |
| Plaque (0°) | 23 | 435 | 500 |
| Plaque (90°) | 23 | 415 | 360 |
| ASTM bar | 60 | 285 | 284 |
| Plaque (0°) | 60 | 289 | 325 |
| Plaque (90°) | 60 | 295 | 250 |

In the case of the ASTM bar specimens (for which the fibres are oriented predominantly along the bar axis, with only a relatively insignificant core of transversely aligned fibres) the predicted failure forces are in good agreement with the experimental results at both 23 °C and 60 °C.

The predicted failure force in flexure for the plaque 0° specimens is approximately 15% less than the experimental value, for both test temperatures. For the plaque 90° specimens, the predicted values at each temperature are approximately 15% above the experimental values. These errors must be, at least partly, due to the significant skin-core structure of fibre orientation in the plaque, as described in Section 7.2.3. Thus, in the 0° direction, the value of $\sigma_y$ will be close to that for the ASTM bar near the surface but will decrease towards the lower bound value as the centre plane is approached. In the 90° direction, the reverse trend will be obtained.

Overall, the above results suggest that an elastic–plastic failure analysis could give reasonable predictions of component failure.

An elastic–plastic failure analysis has been applied to the sections cut from the ribbed plaques discussed in Section 7.6.1. The predicted values of force at failure are compared with the experimental values in Table 7.7. The tensile strength values used in these predictions were derived from tensile tests on specimens cut from the ribs of the plaques (see Table 7.3). The agreement between prediction and experiment in Table 7.7 is considerably better than that obtained using the simple linear elastic analysis in Section 7.6.1 (Table 7.3).

TABLE 7.7
Comparison of Predicted and Experimental Force at Failure for Sections taken from the Ribbed Plaques

Predictions based on a total plastic collapse analysis.

| Material | Rib section | Force at failure (N) | |
|----------|-------------|----------|--------------|
| | | Predicted | Experimental |
| GFPP | 1 | 724 | 681 |
| GFPP | 2 | 548 | 623 |
| GFPP   • | 3 | 385 | 375 |
| GFPA | 1 | 1 740 | 1 483 |
| GFPA | 2 | 1 330 | 1 383 |
| GFPA | 3 | 892 | 817 |

The fibre orientation distribution within the ribs is similar to that found in tensile bars, with a preferred alignment of the fibres in the longitudinal direction. The fibre distribution in the base of the plaque is however closer to random-in-the-plane. The relatively small differences between the predicted and experimental values of failure force in Table 7.7. could be due to the failure to take this difference in fibre orientation (and hence mechanical properties) into account. However, it should also be noted that the tensile stress–strain behaviour of the materials departed significantly from the ideal form given in Fig. 7.20 and total plasticity may not have been attained at component failure.

If lower bound tensile strength values had been used in the above elastic–plastic analysis, the predicted values of failure force would be significantly lower than those given in Table 7.7. Thus all the predicted values would be 'safe' but still in much better agreement with experiment than those values obtained using lower bound data in a simple linear elastic analysis.

## 7.7. CONCLUDING REMARKS

The deformation studies described in Section 7.5 were limited to the case of static loading, i.e. to those situations for which the 'pseudo-elastic' design method was appropriate. The results suggest that the random-in-the-plane fibre orientation distribution (RITP) approach to design for stiffness offers the designer a reasonable compromise between accuracy and ease of use. In this approach, modulus data corresponding to a fibre orientation distribution in which the fibres are oriented randomly in the plane of the moulding are used, and the design calculations are carried out assuming the material to be isotropic and homogeneous.

In many instances, the errors associated with the difficulty of defining, and allowing for, the stress and temperature histories experienced by the component in service will be significantly greater than the error associated with the use of the random-in-the-plane modulus approach.

The reliability of the RITP approach decreases in those instances where the component (or a critical part of a larger component) is of simple form and the stress field is also of simple form (e.g. uniaxial tension). For such cases, the accuracy of the design calculations could be improved, with little or no increase in complexity of the calculations, if the fibre orientation distribution were known. Fortunately, current knowledge of the influence of mould geometry and base polymer on fibre orientation distribution (and hence on stiffness) is probably adequate for reasonable estimates of fibre orientation distribution to be made in these simple cases. Modulus values appropriate to the estimated fibre orientation distribution and the stress field can then be used in the design calculations.

The discussion on design for strength with short fibre reinforced thermoplastics has been limited to the case of static loading. As the fibre orientation distribution throughout a complex component cannot yet be predicted at the design stage, it was suggested that, in critical areas, the possibility of failure should be examined using lower bound tensile strength data for the composite. For cases in which the dominant deformation mode is basically uniaxial tension, the application of this lower bound approach should be straightforward, and predictions based on this approach should be safe (though rather over-pessimistic on occasion).

For those cases in which the dominant deformation mode is flexure, consideration must be given to the choice of an appropriate failure criterion. A simple approach (in common use) is to assume that the component fails when the maximum stress within it reaches the tensile strength of the material, and to relate the applied forces to the stresses in the component via a linear elastic analysis. The strength studies on the ribbed plaque (moulded in GFPP and GFPA) and the gear selector (moulded in GFPP and GFPBT) involved flexural deformation. The results indicate that, for flexure, the use of lower bound tensile strength data with a 'maximum tensile stress' failure criterion in an elastic analysis will lead to excessive over-design in many instances. A possible reason for this is that some degree of plastic collapse occurs prior to complete 'failure' of the component.

The study on the ribbed plaque suggests that the use of lower bound tensile strength data in a plastic collapse analysis will lead to more realistic, but still conservative, predictions of component performance. It is apparent that this approach should be considered only if the material fails in a ductile manner in the tensile tests carried out to establish the design data. Unfortunately, a component moulded in such a material may still fail in a brittle manner in service and further work is needed before reliable guidelines can be given on the range of applicability of this approach.

Irrespective of the type of analysis carried out, lower bound strength data should be used, unless the fibre orientation distribution is known with certainty to be more favourable. Greater understanding of the development of fibre orientation in moulded parts is therefore needed if short fibre reinforced thermoplastics are to achieve their full potential as engineering materials.

## REFERENCES

1. Crawford, R. J., *Plastics Engineering*, Pergamon Press, Oxford, 1981.
2. Williams, J. G., *Stress Analysis of Polymers*, 2nd edn, Longman, London, 1980.
3. Powell, P. C., *Engineering with Polymers*, Chapman and Hall, London, 1983.
4. British Standard BS4618, *Recommendations for the Presentation of Plastics Design Data*, Parts 1.1. and 1.1.1, 1970.
5. Darlington, M. W. and Turner, S., Creep of thermoplastics, in *Creep of Engineering Materials*, C. D. Pomeroy (Ed.), Mechanical Engineering Publications, London, 1978, Ch. 11.
6. Hunt, D. G. and Darlington, M. W., *Polymer*, 1978, **19**, 977.
7. Christie, M. A. and Darlington, M. W., *Plastics and Polymers*, 1975, **43**, 149.
8. McCammond, D. and Benham, P. P., *Plastics and Polymers*, 1969, **37**, 475.
9. Ogorkiewicz, R. M. (Ed.), *Thermoplastics: Properties and Design*, Wiley-Interscience, London, 1974.
10. Marshall, G. P., *Plast. Rubber Proc. Appl.*, 1982, **2**, 169; 1983, **3**, 281.
11. Moore, D. R., Hooley, C. J. and Whale, M., *Plast. Rubber Proc. Appl.*, 1981, **1**, 121.
12. British Standard BS4618, *Recommendations for the Presentation of Plastics Design Data*, Part 1.
13. Titow, V. W. and Lanham, B. J., *Reinforced Thermoplastics*, Applied Science Publishers, London, 1975.
14. Lunt, J. M. and Shorthall, J. B., *Plast. Rubber Processing*, 1980, **5**, 37.
15. Folkes, M. J., *Short Fibre Reinforced Thermoplastics*, Research Studies Press (John Wiley), Chichester, 1982.
16. Herrmann, H., *Polym. Eng. Reviews*, 1983, **2**, 227.
17. Kaliske, G. and Seifert, H., *Plast.-u-Kaut.*, 1975, **22**, 739.
18. Darlington, M. W., McGinley, P. L. and Smith, G. R., *Plast. Rubber Mater. Appl.*, 1977, **2**, 51.
19. Darlington, M. W., Gladwell, B. K. and Smith, G. R., *Polymer*, 1977, **18**, 1269.
20. McNally, D., *Polym. Plast. Technol. Eng.*, 1977, **8**, 101.
21. Bright, P. F. and Darlington, M. W., *Plast. Rubber Proc. Appl.*, 1981, **1**, 139.
22. Bright, P. F. and Darlington, M. W., Preprint No. 4, Plastics and Rubber Institute Conference *Moulding of Polyolefines*, London, 1980.
23. Dunn, C. M. R. and Turner, S., Conference on *Composites — Standards,*

Testing and Design, National Physical Laboratory, UK, April, 1974. (Proceedings published by IPC Science and Technology Press).

24. Hegler, R. P., *Kunstoffe*, 1984, **74**, 271.
25. Darlington, M. W. and Smith, G. R., *Polymer*, 1975, **16**, 459.
26. Darlington, M. W. and McGinley, P. L., *J. Mater. Sci.*, 1975, **10**, 906.
27. Smith, G. R., Darlington, M. W. and McCammond, D., *J. Strain Analysis*, 1978, **13**, 221.
28. Darlington, M. W. and Christie, M. A., in *The Role of the Polymeric Matrix in the Processing and Structural Properties of Composite Materials*, J. C. Seferis and L. Nicolais (Eds), Plenum, New York, 1983, p. 319.
29. Christie, M. A., Darlington, M. W., McCammond, D. and Smith, G. R., *Fibre Sci. Technol.*, 1979, **12**, 167.
30. Akay, G., in *Interrelations Between Processing, Structure and Properties of Polymeric Materials*, J. C. Seferis and P. S. Theocaris (Eds), Elsevier, Amsterdam, 1983.
31. Christie, M. A., Darlington, M. W. and Smith, G. R., Paper 7, *British Plastics Federation Reinforced Plastics Congress, Brighton, UK*. (Published by the BPF, 1978).
32. Stephenson, R. C., Turner, S. and Whale, M., *Composites*, 1979, **10**, 153.
33. Stephenson, R. C., Turner, S. and Whale, M., *Polym. Eng. Sci.*, 1979, **19**, 173.
34. Stephenson, R. C., *Plast. Rubber Mater. Appl.*, 1979, **4**, 45.
35. Darlington, M. W. and Saunders, D. W., in *Structure and Properties of Oriented Polymers*, I. M. Ward (Ed.), Applied Science Publishers, London, 1975, Ch. 10.
36. Roark, R. J. and Young, W. C., *Formulas for Stress and Strain*, 5th edn, McGraw-Hill Kogakusha, Tokyo, 1975, Table 26.
37. Darlington, M. W. and Upperton, P. H., *Plastics and Rubber International*, 1985, **10**(2), 35.

*Chapter 8*

# Coupling and Interfacial Agents and Their Effects on Mechanical Properties

EDWIN P. PLUEDDEMANN

*Dow Corning Corporation, Midland, Michigan, USA*

## 8.1. INTRODUCTION

A wide variety of fibrous and non-fibrous mineral materials are available to enhance properties of thermoplastic products. This enhancement may affect mechanical properties, thermal stability, electrical insulation or conductivity, fire retardance, and cost. Fibers are generally considered as reinforcements and enhance properties at higher cost. Particulate fillers generally reduce costs but can also be reinforcing with proper bonding between polymer and filler. Theory of fiber reinforcement of thermoplastics is discussed in Chapter 2.

The greatest advance made in recent years in filled thermoplastics has been the application of coupling and interfacial agents to enhance the processing and performance of these composites (Table 8.1).

TABLE 8.1
Representative Commercial Coupling Agents and Interfacial Modifiers

| Organofunctional group | Chemical structure |
|---|---|
| A  Vinyl | $CH_2 = CHSi(OCH_3)_3$ |
| B  Chloropropyl | $ClCH_2CH_2CH_2Si(OCH_3)_3$ |
| C  Epoxy | $\overset{O}{\overset{/\backslash}{CH_2CHCH_2OCH_2CH_2CH_2Si(OCH_3)_3}}$ |
| D  Methacrylate | $CH_2 = \overset{\overset{\textstyle CH_3}{\textstyle |}}{C} - COOCH_2CH_2CH_2Si(OCH_3)_3$ |
| E  Primary amine | $H_2NCH_2CH_2CH_2Si(OC_2H_5)_3$ |
| F  Diamine | $H_2NCH_2CH_2NHCH_2CH_2CH_2Si(OCH_3)_3$ |
| G  Mercapto | $HSCH_2CH_2CH_2Si(OCH_3)_3$ |
| H  Cationic styryl | $CH_2{=}CHC_6H_4CH_2NHCH_2CH_2NH(CH_2)_3Si(OCH_3)_3 \cdot HCl$ |
| I  Stearate | $CH_3(CH_2)_{16}COOH$ and salts |
| J  Isostearate | $[(CH_3)_2CH(CH_2)_{14}COO]_3TiOCH(CH_3)_2$ |
| K  Pyrophosphate | $[(C_8H_{17})_2\overset{\overset{\textstyle O}{\textstyle \|}}{P}-O-\overset{\overset{\textstyle O}{\textstyle \|}}{\underset{\underset{\textstyle OH}{\textstyle |}}{P}}-O]_3TiOCH(CH_3)_2$ |

## 8.2. THE TOTAL PICTURE

Applications of 'coupling agents' for surface modification of fillers and reinforcements in plastics have generally been directed towards improved mechanical strength and chemical resistance of composites related to improved adhesion across the interface. Although adhesion is central to any 'coupling' mechanism, it is recognized that many factors are involved in the total performance of a composite system. The interface, or interphase region, between polymer and filler involves a complex interplay of physical and chemical factors related to composite performance as indicated in Fig. 8.1.

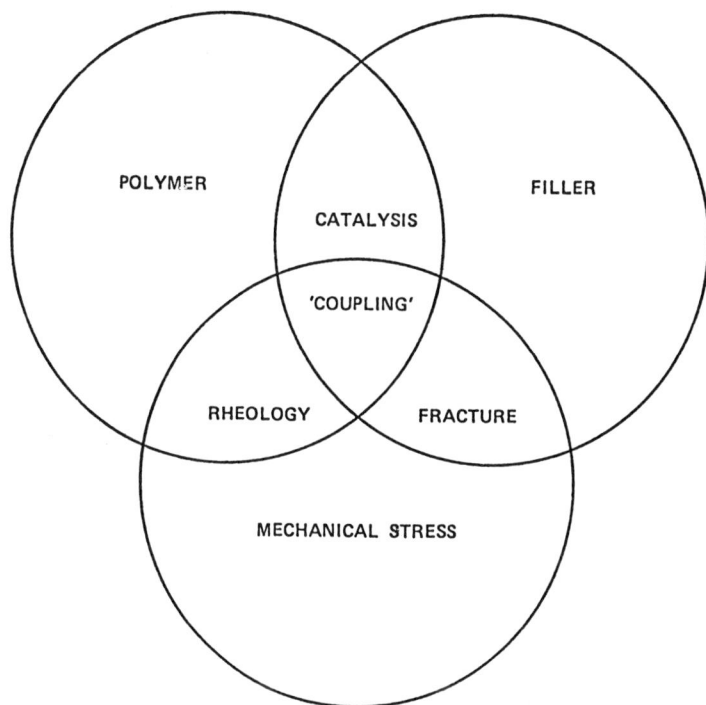

**Fig. 8.1.** Interrelationships of polymer, filler, and mechanical stress in composites.

It is recognized that the total coupling mechanism involves all of these areas and that they are interrelated. Under ideal conditions a treated filler will wet-out and disperse readily in the plastic with Newtonian flow. The treatment protects the filler against abrasion and cleavage during mixing and in the final composite. It also promotes changes in morphology of the polymer at the interface and provides toughness to the interphase region. The treated filler should remain chemically inert to the plastic during mixing but combine with the polymer during the final cure or molding operation.

## 8.3. MATERIALS

Suppliers of organofunctional silanes and titanates generally have disclosed chemical compositions of their materials in order to aid chemists in making intelligent selections of surface modifiers. Standard chemicals such as fatty acids, amides or amines may also be selected from their known chemistry. In addition, many proprietary modifiers of undisclosed composition such as polymeric acids, oils and functional silicones have been recommended for specific applications, but they are difficult to categorize because details of their chemistry are unknown. Some typical materials of known structure are described in Table 8.1.

Filler suppliers often offer both untreated and treated fillers. Some of the suppliers describe the nature and amount of modifier on their fillers, while other filler treatments are held as proprietary. All fiberglass sizes and finishes are proprietary and are described only as being 'compatible' with certain resin systems.

Properties of filled thermoplastics dependent upon properties of the filler itself (shape, particle size, modulus, etc.), or the matrix polymer, are discussed elsewhere (Chapters 3, 4 and 5). The effect of coupling agents and other interfacial agents on properties of thermoplastic composites are discussed in this chapter.

The general subject of silane coupling agents was reviewed recently.[1] Specific applications of silanes as additives for plastics was reviewed by Plueddeman[2] with additional observations on the effect of additives on viscosity of filled polymers.[3] Other coupling agents in thermoplastics, with special emphasis on titanates, were described by Monte and Sugarman.[4] Typical paint additives as hydrophobic wetting agents in filled plastics were described by Cope and Linnert.[5] Organosilicone additives of undisclosed structure were described by Godlewski for modification of highly filled polyolefins.[6]

## 8.4. SCREENING TESTS

It is relatively easy to compare commercial treated fillers with untreated fillers in actual composites to determine whether the advantages of the treatment (ease of fabrication, filler loading, and composite properties) are sufficient to justify the added cost of treated materials. Formulators who plan to treat fillers might desire simple screening tests to determine which treatment, how much treatment, and which method of application might give the best cost-effectiveness.

### 8.4.1. Adhesion

If true 'coupling' of polymer to filler is desired for maximum flexural strength, tensile strength, and retention of properties under humid conditions, a simple adhesion test of polymer to primed glass microscope slides will be informative. Polymers that bond to primed glass will bond to most mineral fillers with comparable treatment. Performance of silane-treated fillers in rubbers correlated well with the degree of improved adhesion in the silanes contributed to films of rubber vulcanized against primed glass.[6] Adhesion of polypropylene fused against primed microscope slides also correlated well with strength properties of injection molded polypropylene filled with wollastonite having similar silane treatments.[7]

### 8.4.2. Wet-out by Liquids

Drops of a series of liquids of varying polarity may be placed in a lightly tamped depression in treated filler and observed for time of complete soak-in. In this way, the effect of change in structure of a series of organofunctional silanes was observed on wet-out of treated glass. Glass microspheres (Potters 3000) were treated with various derivatives of a diamine-functional silane (F in Table 8.1), and tested for wet-out by mineral oil, liquid epoxy, polyester, polyglycol, and water (Table 8.2). Polar aromatic substituents provided a hydrophobic surface with improved wet-out by polyester. A dodecylbenzyl derivative did not wet-out readily by any of the liquids. This indicates that hydrophobic surfaces are not always organophilic, but that the organic group must have good compatibility with the organic resin for maximum wet-out.

### 8.4.3. Viscosity

Wet-out, flow, potential filler loading, etc., can be predicted from simple viscosity measurements of filled liquid polymers. Prototype liquids may be substituted for solid resins, e.g. mineral oil for polyolefins, ester plasticizers for polyesters, etc. A master-mix of filler in liquid may be modified by adding increments of various additives until a minimum viscosity is observed. It is advantageous to let each mix stand for about

TABLE 8.2

Glass Microspheres (Potters 3000) Treated with Silanes (Pick-up about 0·1%)

| R-group in silane $(CH_3O)_3Si(CH_2)_3NHCH_2CH_2NHR$ | Wet-out time (min) | | | | |
|---|---|---|---|---|---|
| | Polyester | Epoxy | Oil | Glycol | $H_2O$ |
| Control (no silane) | 6 | 3 | 3 | 2 | 2 |
| H (unmodified) | 20 | 1 | 1 | 3 | 2 |
| —$CH_2CH_2COOCH_3$ | 14 | 3 | 2 | 3 | 3 |
| —$CH_2C_6H_5$ | 10 | 3 | 2 | 6 | >120 |
| —$CH_2C_6H_4CH{=}CH_2$ | 10 | 3 | 2 | 10 | >120 |
| —$CH_2C_6H_3Cl_2$ | 4 | 5 | 1 | 1 | >120 |
| —$CH_2C_6H_4C_{12}H_{25}$ | 70 | 120 | 14 | 180 | >120 |
| —$CH_2C_6H_4OC_6H_5$ | 12 | 3 | 1 | 2 | >120 |

an hour at room temperature to allow the additive to disperse to the filler surface.

The effect of additives on viscosity of filled polymers showed a predictable pattern based upon acid–base properties of polymer, filler and additive.[3] The effect of 0·4% additive (based on filler) on the viscosities of various filled polyesters are shown in Table 8.3. The polyester was considered to be weakly acidic. Similar data were obtained with neutral and basic polymers.

(a) Neutral polymers required surface-active additives with all fillers for good dispersion. Almost any polar additive will lower viscosity markedly, but acid-functional additives are most effective with basic fillers, and basic additives are recommended with acid fillers.

(b) Acid fillers in basic polymers or basic fillers in acid polymers give fairly good dispersions without additives. Lewis acids (titanates and aluminum alkoxides) may be beneficial on acid fillers in basic polymers, but should not be used with basic fillers in acid polymers.

(c) Additives may be very helpful in dispersing acid fillers in acid polymers or basic fillers in basic polymers. Cationic silanes or Lewis acids are of most benefit on acid fillers in acid polymers, and may be of some benefit on basic fillers in basic polymers.

(d) Neutral silanes (silanes A–D of Table 8.1) modified with catalytic amounts of an amine or a titanate are generally more effective as

TABLE 8.3
Percentage Viscosity Change in Filled Polyester (Paraplex P-13)

| 0·4% Additive based on filler | Filler | | | | |
|---|---|---|---|---|---|
| | 50% Silica | 33% Clay | 29% Talc | 64% Al(OH)₃ | 67% CaCO₃ |
| Control (Nsm$^{-2}$) | 22·0 | 48·0 | 16·0 | 14·5 | 12·9 |
| *Organic additives* | | | | | |
| Undecanoic acid | −20 | −3 | −5 | +5 | +87 |
| 1-Decanol | −32 | −33 | −23 | +23 | −19 |
| 1-Hexylamine | −89 | −96 | −47 | +110 | +100 |
| Ti(OBu)₄ | −73 | −86 | −34 | +400 | +100 |
| Al(OBu)₃ | −77 | −50 | −33 | +200 | +100 |
| *Silane additives* | | | | | |
| Silane D (methoxy) | −19 | −7 | −15 | −12 | −16 |
| Prehydrolyzed D (silanol) | −30 | −33 | −4 | −21 | −27 |
| Silane D–Ti(OBu)₄ catalyst | −39 | −59 | −22 | +69 | +72 |
| Silane H | −83 | −92 | −31 | +55 | +29 |

additives than the pure silane in modifying viscosity. Performance as a coupling agent also is improved by such modification.

Surface-active additives that reduce viscosity of filled polymers do not necessarily improve the properties of composites unless they also improve adhesion of the polymer to filler (Table 8.4). Silane D (Table 8.1) had little effect on viscosity, but was a good coupling agent, while Silane H reduced viscosity markedly and also gave good coupling. Organic titanates were effective viscosity reducers but gave no coupling.

### 8.4.4. Daniel's Flow Point Test
A simple test is needed to determine whether the surface modifier has been properly dispersed on the filler surface. It is possible to get very poor results with very good materials if they are not applied properly.

A simple means of determining the relative dispersibility of treated fillers in liquid resins is the Daniel's flow point test, which consists of adding liquid vehicle (15–25% resin in a neutral solvent) from a buret to a known amount of dry pigment while stirring with a spatula. A first

TABLE 8.4
Properties of Polyester (P-43) with Silica Filler (50% Minusil: 5 μm)

| 0·4% Additive based on filler | Viscosity of mix (Nsm$^{-2}$) | Flexural strength of castings (MPa) | |
|---|---|---|---|
| | | Dry | Wet (2 h boil) |
| None | 22·0 | 115 | 70 |
| Silane D | 21·5 | 163 | 139 |
| 10% TBT in silane D | 20·0 | 178 | 151 |
| TBT[a] | 16·6 | 106 | 75 |
| TTM-33[b] | 10·0 | 135 | 72 |
| Silane H | 8·5 | 184 | 130 |

[a]TBT, tetrabutyl titanate.
[b]TTM-33, Kenrich isopropyl trimethacryl titanate.

end-point (wet point) occurs when the pigment barely sticks together to form a ball. The second (flow point) is observed when the mixture becomes fluid enough to flow off the spatula, and the last drop falling pulls with it a string which breaks and snaps back. (Highly thixotropic mixes do not give sharp end-points.)

The wet point measures the amount of liquid necessary to wet the pigment but not to disperse it. The flow point measures the amount of liquid necessary to fill the interstices between the pigment particles plus an amount necessary to provide a slick enough coating around each pigment particle to deflocculate it.

A mineral oil (No. 2 fuel oil) has been used as vehicle to predict the viscosity of treated fillers in molten polyolefins.

The Daniel's flow point test was used to compare different silane treatments on silica to be used as a filler in polyester composites.[2] In this test, 6 g of filler was titrated with a 25% solution of Paraplex P-43 in styrene (Table 8.5). The silica filler was treated by dry-blending with the two best-known coupling agents (silanes D and H in Table 8.1) for polyester resins. Small amounts of polar solvents were added to aid in dispersing the silane over the silica surface. The wet point did not vary much among the silane-treated silicas, but the flow point was a better indication of the uniformity of silane treatment on the filler (Table 8.5). These data suggest that a good coupling agent on a filler can give much less than optimum performane in a composite. Incomplete dispersion of the coupling agent over the filler surface shows itself in high flow point titration and in low mechanical performance of composites. It

TABLE 8.5
50% Treated Silica in Polyester Castings

| | | *Flow* | *Flexural strength of castings (psi)* | |
|---|---|---|---|---|
| *Treated silica* | *Solvent for silane* | *point (ml)* | *Dry* | *Wet (2 h boil)* |
| 0·3% Silane D and 3% solvent on Minusil 5 μm | Water | 5·5 | 20 000 | 15 800 |
| | Methanol | 5·9 | 20 300 | 15 700 |
| | No solvent | 3·9 | 24 900 | 24 900 |
| | *n*-Butanol | 5·0 | 24 900 | 20 900 |
| | 2-Ethylhexanol | 4·2 | 25 700 | 24 900 |
| | Untreated | 7·0 | 14 900 | 9 400 |
| 0·3% Silane H and 3% solvent on Minusil 5 μm | No solvent | 5·3 | 23 900 | 13 000 |
| | Isopropanol | 5·0 | 26 000 | 18 000 |
| | Dowanol EM[a] | 3·7 | 24 000 | 21 200 |

[a]Ethylene glycol monomethyl ether.

also suggests that a non-polar silane D does not require a polar solvent to aid dispersion on a filler, but that a cationic silane H does not disperse well on the filler by dry-blending unless a polar solvent is added as a carrier. It is recommended that any silane treatment of fillers for commercial application in filled plastics be monitored by a Daniel's flow point test.

## 8.5. MECHANISM OF ADHESION

Water-resistant adhesion of organic polymers to hydrophilic mineral surfaces requires chemical bonds across the interface. Dispersion forces or electrostatic attraction may contribute towards initial adhesion, but these cannot prevent loss of adhesion due to intrusion of water.

Many polar organic functional groups will form covalent oxane bonds with mineral oxide surfaces, but all of such bonds are readily hydrolyzable. Adhesion promoters such as certain compounds of silicon, titanium, chromium, and phosphorus also form oxane bonds with mineral oxide surfaces. Although they may be somewhat more stable than organic oxane bonds, they also are hydrolyzable. Formation and hydrolysis of oxane bonds is a reversible reaction, such that the advantage of inorganic 'coupling agents' may be a more favorable equilibrium constant for oxane bond formation.

Since all bonds between polymer and mineral are ultimately

hydrolyzable, the adhesion promoter must also provide mechanical properties to the interface region that retain functional groups in contact with the surface so that equilibrium conditions are maintained. Silane coupling agents are uniquely suited to provide a favorable equilibrium constant for oxane formation with the added capability to crosslink at the interface for optimum mechanical properties.[1]

Silane coupling agents are deposited on mineral surfaces as multi-molecular layers of oligomeric siloxanes in a soluble state. When brought into contact with an organic polymer, the coupling agent may provide adhesion through several mechanisms.

(a) *The oligomeric siloxane* is compatible with the polymer and crosslinks with it during cure. This is the primary mechanism of bonding thermosetting resins and vulcanizable elastomers.

(b) *The oligomeric siloxane* is partially compatible with the polymer and crosslinks into an interpenetrating polymer network. This is the predominant mechanism with thermoplastic resins and rubbers. When only the siloxane layer crosslinks, a pseudo-interpenetrating polymer network is produced.

(c) *A silane modified polymer primer* may interdiffuse into a thermo-plastic polymer and bond by chain entanglement. Solution compatibility is of prime importance, although some crosslinking of the primer may also be required.

## 8.6. EFFECT OF COUPLING AGENTS ON COMPOSITE PROPERTIES

Many surface-active agents such as the stearates, organic hydrophobic wetting agents, organic titanates, and non-reactive silanes may be very effective in protecting a mineral surface (especially glass fibers) against abrasion and fracture, and in improving wet-out and dispersion in the resin and the rheology of the melt during fabrication without providing true coupling across the interface. Their major value is their contribution toward improved processing (see Chapter 5) and the possibility of higher filler loading. Surface-active agents may act as release agents between filler and polymer, which will generally result in increased impact strength and elongation, but decreased tensile, flexural and shear strengths.

Some properties of composites such as density and modulus are straight-line relationships with filler loading and may be estimated from a simple rule of mixtures. Surface-active agents may have a minor influence on these properties by eliminating voids in a composite, or they may allow higher filler loadings to obtain higher density and modulus.

Other properties may be strongly dependent upon adhesion between filler and polymer.

### 8.6.1. Shear Strength

The shear strength of a composite is directly related to the adhesion between polymer and reinforcement. Measurement of adhesion is a complex problem, since values observed in a debonding experiment are strongly dependent upon the mode of failure. Numerous tests have been devised to measure the stress-transfer capability between resins and reinforcements. Although these may not be measurements of true adhesion, they are informative in showing the effect of resin properties and fiber finishes on resistance to debonding.

In an ingenious test described by Frazer *et al.*,[9] a single fiber encapsulated in a tensile test specimen is subjected to traction to an elongation greater than that required to fracture the fiber. When the plastic matrix deforms beyond the elongation-to-break of the fiber, the fiber will break at its weakest point. One now has two pieces of fiber of different lengths. Under continued deformation, the fiber stress continues to increase until it reaches the breaking stress of one of the two fragments, at which point a second fracture occurs. This process will continue until all the fragments are smaller than their critical lengths, after which time fiber fracture is no longer possible. If the tensile strength of the fiber is known and is independent of fiber length, one may calculate an effective interfacial shear strength from the measured critical length.

Single fibers of E-glass with an aminofunctional silane size were pulled in nylon 6, polypropylene, and a carboxy-modified polypropylene. The cumulative distributions of fragments from ten specimens of each system indicated an effective interfacial strength of 34·4 MPa for nylon 6, 7·1 MPa for polypropylene, and 21·0 MPa for the carboxy-modified polypropylene. These results are qualitatively in agreement with observations on simple microscope slide adhesion tests.

### 8.6.2. Impact Strength

Impact strength of a composite does not have a simple relationship with adhesion between filler and polymer. The effect of adhesion at the glass–resin interface on impact strength of the composite was studied by comparing short-beam shear tests of glass fiber reinforced epoxies and polyesters with cantilever cleavage tests and Charpy impact tests on similar composites.[10] Highest total impact strength corresponded to poorest adhesion resulting from a release agent on the reinforcement. Somewhat better adhesion obtained with clean glass gave a minimum impact strength, which increased again with good coupling agents on the glass surface (Fig. 8.2). Improved impact strength, therefore, cannot be used as an indication of coupling at the interface, since it could also indicate adhesion or release properties.

Alesi and Barron[11] described a composite body armor with a ceramic facing, backed by a fiberglass reinforced laminate capable of

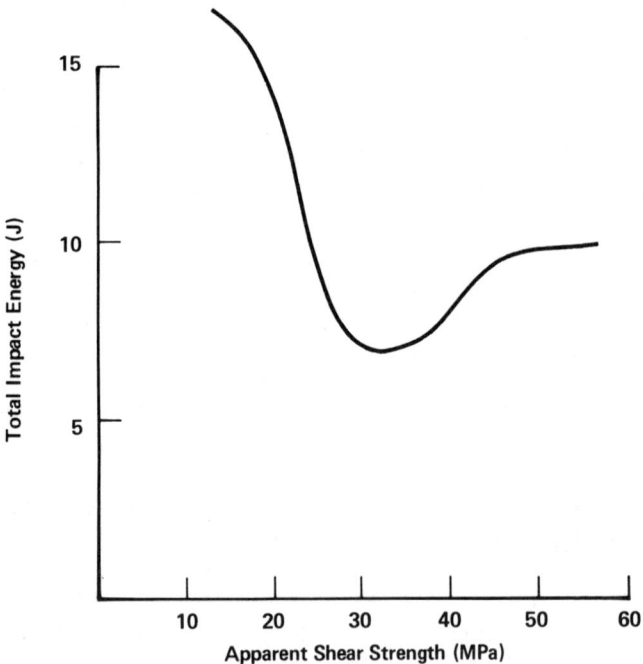

**Fig. 8.2.** Impact energy versus short-beam shear strength for fiberglass reinforced polyester laminates.

delaminating over a relatively large area at the point of impact. Resin-glass adhesion was designed to be poor, but adequate to withstand mechanical (as opposed to ballistic) impact.

### 8.6.3. Flexural and Tensile Strengths

A flexural test subjects the interface to a complex mixture of tension, compression and shear such that results are difficult to interpret in terms of mechanics, but the test is simple to run and relates well to composite performance.

Tensile tests may be more a function of reinforcement alignment than of adhesion across the interface. Tensile and flexural strengths are generally determined on injection molded filled thermoplastics.

Gaehde[12] reported on mechanical properties of clay filled poly-ethylene with three silane treatments on the clay. Although vinylsilanes graft to polyethylene while methacrylate silanes homopolymerize at the interface, the methacrylate silane was most effective in increasing the crystallinity of polyethylene at the interface. Homopolymerization and self-condensation of methacrylate- and amine-functional siloxanes in contact with molten polymer may also have trapped polyethylene in an interpenetrating polymer network (IPN), at the interface. Mechanical properties of the kaolin filled polyethylene seemed to be more dependent upon polymer structure at the interface than upon chemical grafting (Table 8.6).

TABLE 8.6
Silane-treated Kaolin in High-density Polyethylene

| Silane on kaolin | Mode of action | Mechanical strength of composites (20 vol. % clay) | | |
|---|---|---|---|---|
| | | Tensile (MPa) | Flexural (MPa) | Impact (J) |
| None | — | 19 | 37 | 13·7 |
| Silane A | Graft | 20 | 38 | 13·7 |
| Silane D | Orient + IPN | 24 | 49 | 16·4 |
| Silane E | IPN | 22 | 41 | 15·0 |

An interpenetrating polymer network (IPN) was also proposed for coupling particulate fillers to polypropylene.[8] Addition of a trace of dicumyl peroxide to silane H aided in crosslinking the interphase region. Chlorinated paraffin added to polypropylene improved compatibility of the polymer and filler and may have contributed additional crosslinking by reacting with the silane (Table 8.7).

TABLE 8.7
Polypropylene with 35% Wollastonite[a] Filler

| Coupling system with filler | Flexural strength (MPa) | Tensile strength (MPa) | HDT (°C) at 1·8 MPa | Falling dart impact (J) |
|---|---|---|---|---|
| No filler | 60 | 33 | 56 | 2·8 |
| No silane | 58 | 28 | 90 | 6·7 |
| SilaneD | 57 | 29 | — | 6·0 |
| D + P[b] + CP[c] | 59 | 30 | — | 5·6 |
| Silane F | 58 | 28 | — | 6·7 |
| F + CP | 62 | 34 | — | 5·6 |
| Silane H | 60 | 29 | — | 6·7 |
| H + P | 65 | 33 | 118 | — |
| H + P + CP | 68 | 33 | 118 | 6·7 |

[a]Nyad-G (Nyco Div.) in Profax 6223 (Hercules).
[b]1% Dicumyl peroxide added to silane.
[c]1% Chlorez® 700 (Dover Chem.) and 0·05% MgO added to PP.

Improvement in flexural strength imparted by silanes in thermoplastic composites is generally greater when test specimens are compression molded than when they are injection molded. Poorer performance in injection molding suggests that structures in the interphase region promoting adhesion may be destroyed by high shear during composite formation (Table 8.8).

Flexural strengths of four-ply fiberglass cloth laminates prepared by compression molding are listed in Table 8.9. Silane treatments on glass were generally beneficial. Silane H was better than silane F. Addition of dicumyl peroxide to silane H did not always give additional improvement. It is probable the peroxide-modified silane finish on glass might provide optimum performance at lower molding temperatures.

### 8.6.4. Heat Distortion Temperature

Addition of mineral filler generally raises the heat distortion temperature (HDT) of thermoplastics. The HDT is raised even more with a coupling agent on the filler.[8] Best coupling agents, as indicated by adhesion tests, raise HDT the most (Table 8.10).

TABLE 8.8
Improvement Imparted by Silanes in Glass–Thermoplastic Composites

| Polymer–silane system (Table 8.1) | Flexural strength improvement (%) | | | |
|---|---|---|---|---|
| | Compression molded | | Injection molded | |
| | Dry | Wet | Dry | Wet |
| Nylon–aminosilane F | 55 | 115 | 40 | 36 |
| Nylon–cationic silane H | 85 | 133 | 40 | 45 |
| PBT–aminosilane F | 21 | — | 23 | 24 |
| PBT–cationic silane H | 60 | 47 | 28 | 11 |
| Polypropylene–silane F | 8 | 18 | 7 | 10 |
| Polypropylene–silane H | 86 | 89 | 16 | 16 |

TABLE 8.9
Flexural Strengths (MPa) of Compression Molded Glass Laminates

| Polymer | Silane on glass | | | | | | | |
|---|---|---|---|---|---|---|---|---|
| | None | | Silane F | | Silane H | | Silane H + peroxide | |
| | Dry | Wet[a] | Dry | Wet | Dry | Wet | Dry | Wet |
| Nylon | 218 | 96 | 340 | 228 | 350 | 248 | 405 | 283 |
| Polycarbonate | 250 | 86 | 343 | 211 | 378 | 277 | 355 | 240 |
| Polyterephthalate | 211 | 137 | 250 | 136 | 285 | 172 | 337 | 202 |
| Polystyrene | 242 | 113 | 320 | 222 | 348 | 258 | 382 | 285 |
| SAN | 248 | 153 | 232 | 196 | 334 | 279 | 364 | 287 |
| Polypropylene | 100 | 86 | 112 | 97 | 188 | 157 | 133 | 122 |
| PVC (rigid) | 131 | 93 | 167 | 119 | 225 | 157 | 188 | 132 |

[a]Wet = after 2 hour water boil.

## 8.6.5. Electrical Properties

Retention of electrical properties under wet conditions is one of the most sensitive tests for coupling across the interface. Composites with coupling agents that provide maximum strength properties and heat distortion temperatures generally have best retention of electrical properties. Surface-active agents that lubricate the interface for

TABLE 8.10
Heat Distortion Temperature (HDT) of Polypropylene with 35% Filler
(°C at 1·8 MPa [264 psi] stress)

| Coupling system | Filler | | | | |
|---|---|---|---|---|---|
| | Glass microbeads[a] | Wollastonite | | Suzerite Mica[c] | Talc[d] |
| | | Nyad-400[b] | Nyad-G[b] | | |
| No filler | 56 | 56 | 56 | 56 | 56 |
| No treatment | 63 | 80 | 89 | 78 | 80 |
| Silane H + peroxide | 70 | 84 | 118 | 85 | 87 |
| Silane H + peroxide and Chlorez | 69 | 92 | 118 | 89 | 88 |

[a]Potters 3000, PQ Corp.
[b]Nyco Div. of Processed Minerals, Inc.
[c]Suzerite 325, Martin Marietta.
[d]Pfizer MP 50-35.

maximum impact strength will not provide good wet electrical properties. Continuous fibrous reinforcements especially require good coupling because of the wicking action of water along fiber surfaces, but even particulate fillers require a coupling agent for retention of wet electrical properties (Table 8.11).

TABLE 8.11
Electrical Properties of Mineral Filled[a] Polyethylene

| Silane[b] added (Table 8.1) | Clay | | | | Wollastonite | | | |
|---|---|---|---|---|---|---|---|---|
| | Dielectric constant | | Dissipation factor | | Dielectric constant | | Dissipation factor | |
| | Dry | Wet | Dry | Wet | Dry | Wet | Dry | Wet |
| None | 2·7 | 3·0 | 0·003 | 0·082 | 2·8 | 4·2 | 0·009 | 0·147 |
| Silane D | 2·7 | 2·7 | 0·002 | 0·003 | 2·8 | 2·9 | 0·007 | 0·014 |
| Silane C | 2·7 | 2·7 | 0·002 | 0·005 | 2·8 | 2·9 | 0·007 | 0·013 |

[a]50 parts filler and 50 parts polyethylene. Electrical properties at 1000 Hz per ASTM D-150.
[b]Silane added to clay at 2 phf (parts per hundred of filler) and to wollastonite at 0·8 phf.

## 8.6.6. Graded Interfaces

Optimum mechanical performance of a mineral reinforced organic polymer composite imposes contradictory requirements on the interface between the polymer and the mineral.

*Optimum stress transfer* between a high-modulus filler and a lower-modulus resin requires an interphase region of intermediate modulus.

*Composite toughness* and ability to withstand differential thermal shrinkage between polymer and filler require a flexible boundary region or a controlled fiber pullout to relieve localized stresses.

Fallick *et al.*[13] described a family of mineral reinforced thermoplastics called Ceraplasts in which individual mineral particles were encapsulated in a thin layer of thermosetting resin having modulus and strength characteristics intermediate between those of filler and polymer matrix. The encapsulating phase must bond firmly to both the mineral and the matrix polymer (e.g. polyethylene).

In an effort to improve impact strength of filled thermoplastics while retaining the mechanical and chemical properties of rigid polymers, Plueddemann[14] encapsulated fillers in a silane-modified rubber. The interface between rubber and mineral was a crosslinked siloxane that contributed strength and chemical resistance. This concept was successful in increasing the impact strength of compression molded polypropylene composites, but failed in injection molded parts, due to high shear that disrupted intermediate layers.

A comparable rubber-modified sandwich was proposed[1] in preparing metal sandwiches with thermoplastic cores. A silane-bonded elastomer layer was beneficial in absorbing interfacial stress caused by a mismatch of thermal expansion between surface layers and core.

## 8.7. CONCLUSION

Many benefits in processing and properties of mineral filled thermoplastics result from a trace of coupling or interfacial agent at the interface. Adhesion across the interface is necessary for improved mechanical strength, maximum heat distortion temperature and wet electrical properties. Two silane primers have been effective in bonding virtually all thermoplastics to glass and should be considered in preparing any thermoplastic composites.

A primer based on silane H containing 1% added dicumyl peroxide is especially recommended for polyolefins and some engineering thermoplastics. A primer comprising a cold blend of 10 parts of silane C in 90 parts of commercial melamine resins is very effective with polar polymers such as the engineering thermoplastics. These and other silanes have performed well with filled thermoplastics as indicated in Table 8.12.

TABLE 8.12
Recommended Coupling Agents for Thermoplastics

| *Thermoplastic* | *Recommended silane (Table 8.1)* |
| --- | --- |
| Nylon | B, C, E, F, H |
| Polyester | C + Melamine, E, F |
| Polycarbonate | C + Melamine, E, F |
| Cellulosic | C |
| PVC plastisol | F, H |
| Polymerizable plastisol | D, H |
| Phenoxy | C + Melamine |
| Polyolefin | H + DiCup |
| Polystyrene | H + DiCup |
| EVA | H |
| Crosslinkable EVA | D + Amine |
| Poly(acrylic ester) | F, H, C + Melamine |
| Acrylic acid copolymer | E, F, C |

# REFERENCES

1. Plueddemann, E. P., *Silane Coupling Agents*, Plenum Press, New York, 1982.
2. Plueddemann, E. P., in *Additives for Plastics*, Vol. 1, R. B. Seymour (Ed.), Academic Press, New York, 1978, pp. 123–67.
3. Plueddemann, E. P., in *Additives for Plastics*, Vol. 2, R. B. Seymour (Ed.), Academic Press, New York, 1978, pp. 49–61.
4. Monte, S. J. and Sugarman, G., in *Additives for Plastics*, Vol. 2, R. B. Seymour (Ed.), Academic Press, New York, 1978, pp. 63–74.
5. Cope, D. E. and Linnert, E., *Preprints, SPI Conf. Reinf. Plast.*, 1980, **35**, 20-F.
6. Godlewski, R. E., *Preprints, SPI Conf. Reinf. Plast.*, 1982, **37**, 6-F.
7. Plueddemann, E. P., in *Developments in Rubber Technology*, A. Whelan and K. S. Lee (Eds), Applied Science Publishers, London, 1980, Ch. 5.
8. Plueddemann, E. P. and Stark, G. L., *Preprints, SPI Conf. Reinf. Plast.*, 1980, **35**, 20-B.
9. Frazer, W. A., Ancker, F. H. and Dibenedetto, A. T., *Preprints, SPI Conf. Reinf. Plast.*, 1975, **30**, 22-A.

10. Yeung, P. and Broutman, L. L., *Preprints, SPI Conf. Reinf. Plast.,* 1977, **32,** 9-B.
11. Alesi, A. L. and Barron, E. R., *Preprints, SPI Conf. Reinf. Plast.,* 1968, **23,** 3-A.
12. Gaehde, J., *Plaste u Kautschuk,* 1975, **22**(8), 626.
13. Fallick, G. L., Bixler, H. J., Marsella, R. A., Garner, F. R. and Fettes, E. M., *Preprints, SPI Conf. Reinf. Plast.,* 1967, **22,** 17-E.
14. Plueddemann, E. P. and Stark, G. L., *Adhesion Soc. Annual Meeting, Savannah, GA,* 1980.

*Chapter 9*

# Fibre–Matrix Interactions in Reinforced Thermoplastics

R. H. BURTON and M. J. FOLKES

*Department of Materials Technology, Brunel University, Uxbridge, UK*

## 9.1. INTRODUCTION

The mechanical properties of fibre reinforced materials depend critically upon the fibre length distribution, fibre orientation distribution and interfacial shear strength. In thermosetting polymers, the last of these is controlled to a great extent by chemical bonding occurring between the fibres and matrix. Furthermore, when considering reinforced thermosets, the simplifying assumption is usually made, explicitly or otherwise, that the polymer matrix remains basically unaffected by the introduction of the reinforcing phase. When using thermoplastics as matrix materials, however, there is no reason to expect that such an assumption will be valid. Indeed, changes taking place at the molecular level in the matrix can have an appreciable influence on the overall properties of the composite. This arises because of the direct effect of the fibres on the processes of crystallisation occurring in the matrix. Special microstructural features may then develop, some of which are only consequent on the presence of fibre surfaces while others rely on the occurrence of melt flow prior to crystallisation. As we will see in this chapter, the interaction between the fibres and the matrix is very complex, but a comprehensive understanding of this is necessary for the rational development of improved composite formulations of commercial significance.

## 9.2. POLYMER CRYSTALLISATION AND NUCLEATION

### 9.2.1. Principal Aspects of Polymer Crystallisation

Thermoplastics can be classified, in general, into crystalline and amorphous polymers. In this section we shall be concerned with crystalline polymers or, more precisely, semicrystalline polymers, since these exhibit, in fact, much less than 90% crystallinity. Under favourable thermal conditions the long-chain molecules composing these materials will line up to produce local areas having an ordered crystalline structure. This propensity to crystallise depends on the molecular conformation. Molecules with bulky side groups will tend to remain disordered, i.e. in an amorphous state. On the other hand, simple long molecules such as polypropylene will readily fall into crystallographic register. It should be noted in passing, however, that some polymers are available in different forms which, although chemically identical, have different stereoregularity; polyethylene is an important example, available in linear and non-linear forms. With other polymers such as polystyrene the criterion is the distribution of the side groups, e.g. isotactic and atactic conformations; the former, being regular, crystallises fairly readily, whereas the latter (the commonly used type) is almost completely amorphous.

The formation of crystal structures is a complex process, and it is only during the last 30 years or so that we have built up our knowledge, incomplete though it still is, of the mechanisms involved. The first step in the growth of such a structure occurs when the chains become attached to a nucleus of some kind, and then fold back regularly on themselves to form plate-like structures, known as lamellae, which may be typically

0·2 μm across (Fig. 9.1). These lamellae then, in turn, may line up to form a specific, larger structure. In many polymers this usually takes the form of spherulites (Fig. 9.2), but other forms, such as shish-kebabs, are also found under certain conditions and we will be discussing these later.

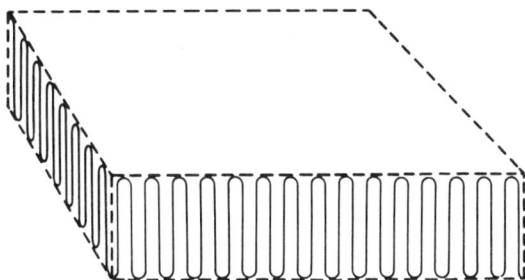

**Fig. 9.1.** Schematic diagram showing the formation of a lamella by folded-chain molecules.[14]

**Fig. 9.2.** Optical micrograph showing spherulites in polypropylene.

In a spherulite, the lamellae radiate in all directions from a nucleus with the lamellar planes parallel to the radius, although in varying directions perpendicular to the radius. The actual molecular chains, however, always lie tangential to the growing spherical front, i.e. at right-angles to the radii. As a result of the orientation of the molecules in the lamellae, the material is locally birefringent. In fact the birefringence is usually defined with respect to the radial direction, i.e.

$$\Delta n = n_r - n_p$$

where $n_r$ and $n_p$ are the refractive indices parallel and perpendicular to the radii, respectively. The birefringence is usually negative, but under certain thermal conditions positive birefringence is encountered in, for example, nylon 66 and polyethylene terephthalate (Terylene).[1–5]

The nature of the nuclei originating such structures is still a matter of doubt, but it is generally believed that in the bulk of the polymer they are usually foreign bodies, e.g. catalyst residues, antioxidants, fillers, or specific nucleating agents, although some nucleation may originate in chance tangling or ordering of polymer chains. The ultimate size of spherulites is governed largely by the density of nucleating sites; the presence of a great number of active nuclei prevents the individual spherulites from growing large.

In cooling from the melt, a temperature is reached at which the polymer chains become attached to the nuclei; above this temperature the molecular movement is too great for this to occur, whereas a low temperature can inhibit attachment, due to too little movement. Similarly there is an optimum temperature, or rather temperature band, for the maximum rate of growth in the folded chain conformation.

Chatterjee et al.[6] and also Campbell and Qayyum[7] have shown that at any temperature only a certain number of potential nuclei are active, and that the number of active nuclei increases as the temperature is lowered. Although they only investigated a limited temperature range covering the region of increasing nucleation it is possible for some polymers to be quenched rapidly down to a temperature at which virtually no nucleation or growth occurs, i.e. they remain essentially amorphous. However, in practice with many polymers it is not possible to quench sufficiently rapidly to avoid nucleation occurring before the lower temperature is reached. For large spherulites to form it is thus necessary to hold the polymer at a relatively high temperature where there is little nucleation, thus minimising the number of nucleating sites, and hence allowing maximum growth to occur without constraint by neighbouring spherulites.

There is considerable disagreement regarding the influence of the size of spherulites on the physical properties of the polymer. This is probably due to the varying influence of competing factors; whereas the spherulites themselves represent stiff, strong material the boundary layer between the spherulites consists of short-chain debris, and is thus a source of weakness. Way *et al.*[8] have shown, in fact, that for polypropylene mouldings the graph of yield stress versus spherulite size goes through a maximum at a critical spherulite size. Scanning electron microscopy (SEM) and optical microscopy indicated that this represented a changeover from spherulitic yield to boundary yield.

### 9.2.2. The Origins of Nucleation

As mentioned above, the precise nature of nucleating sites is still a matter of some controversy. For example, it is possible that nucleation may be due to more than one mechanism, e.g. chemical in some cases, topological in others. Certainly it appears that there is no universally strong nucleating agent, but rather some kind of specific matching between a particular matrix and substrate. Many of these hypotheses refer to nucleation at fibre surfaces, which we shall be dealing with in detail later, but the mechanism, whatever it is, is almost certainly the same as for nucleation by particles.

Turnbull and Vonnegut[9] have postulated a matching between unit cells of the filler and the matrix, but there is much contrary evidence;[7] the unit cell parameters for both nylon 66 and Terylene (polyethylene terephthalate) differ considerably from those of polypropylene, yet the latter crystallises strongly on to both the former substrates.

Chatterjee *et al.*[6] concluded that crystallinity is a necessary, though not sufficient, condition for a substrate to be a strong nucleating agent. Kevlar fibre, however, is strongly nucleating[10] and yet according to Pruneda *et al.*[11] has a non-crystalline skin. Chemical similarity also does not stand up as a generally applicable explanation in that, for example, again essentially non-polar polypropylene crystallises on to polar materials such as nylon, Terylene and Kevlar. Similarly, both low-energy surfaces such as Terylene and high-energy surfaces such as graphite fibres are capable of nucleating crystalline growth. Kontsky *et al.*[12] have also shown that changing the surface energy of substrates by a factor of more than four did not significantly change the nucleating ability of alkali halide single crystals in polyethylene.

Binsbergen,[13] who has carried out a considerable amount of work in the field of heteronucleation, has examined various hypotheses that have been put forward to explain the activity of nucleating agents. These proposed conditions and mechanisms include, as mentioned above, the high surface energy of the substrate, epitaxy and a minimum particle size just compatible with the thickness of the crystallising material in holes on the nucleating particles. However, after rejecting all these possible mechanisms, he concludes that the relevant mechanism is probably nucleation at steps of limited length ('ditches') in the surface of nucleating particles causing pre-alignment of polymer chains, thus facilitating crystallisation.

Although they do not propose any new explanations for the nature of nucleation, Campbell and Qayyum[7] have recently surveyed the evidence for the various hypotheses and also report some interesting work of their own.

Suffice it then to note the subject of nucleation is still at the stage of much conjecture and active investigation.

The nucleation phenomena discussed above refer to those cases where nucleating agents have been deliberately added to the polymer in order to control spherulite size, etc. However, nucleation may arise as a direct consequence of melt flow occurring prior to crystallisation. One well-known example arises when the flow field is primarily extensional. In this case, extended chains will form, which can act as a backbone for the nucleation of the more usual chain-folded crystals. The morphological features so formed are known as shish-kebabs,[14] which are shown schematically in Fig. 9.3. They consist of a string of platelets apparently connected by a central backbone. The molecular direction is along the shish-kebab axis. With such an extended-chain crystalline structure the modulus along the axis is very high, with corresponding low extensibility.

## 9.3. TRANSCRYSTALLINITY

As mentioned previously, some fibres also provide nucleating sites on their surfaces. Isolated spherulites may then be seen to be attached to the fibres; but where the nucleation density is sufficiently high along the fibre surface, the growth takes the form of a sheath surrounding the fibre, and is then known as transcrystallinity (Figs. 9.4 and 9.5). This columnar growth consists, in fact, of embryonic spherulites which have been constrained to grow predominantly in one dimension, rather than three, due to the fact that they are prevented by their neighbours from spreading out sideways, and hence can grow only normal to the surface. Keller[15] confirmed, however, that the structure is otherwise identical to

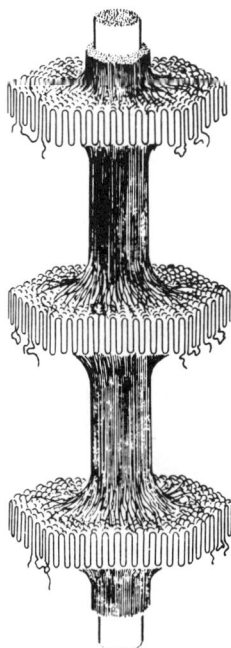

**Fig. 9.3.** Schematic diagram showing the structure of a shish-kebab.[44]

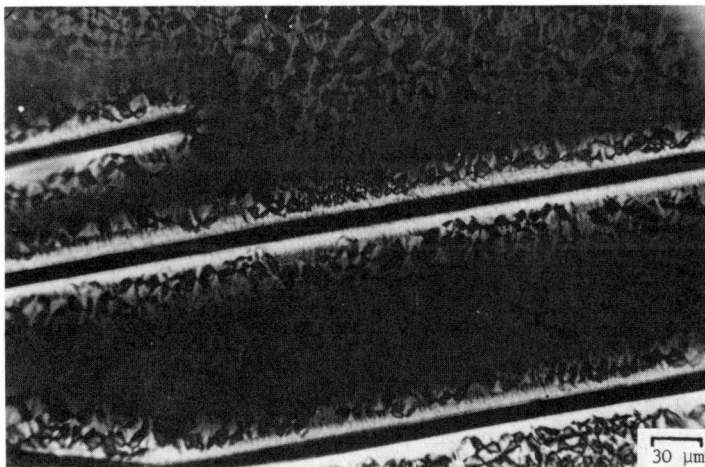

**Fig. 9.4.** Optical micrograph showing transcrystalline growth around HMS (Type I) carbon fibres in nylon 66.[16]

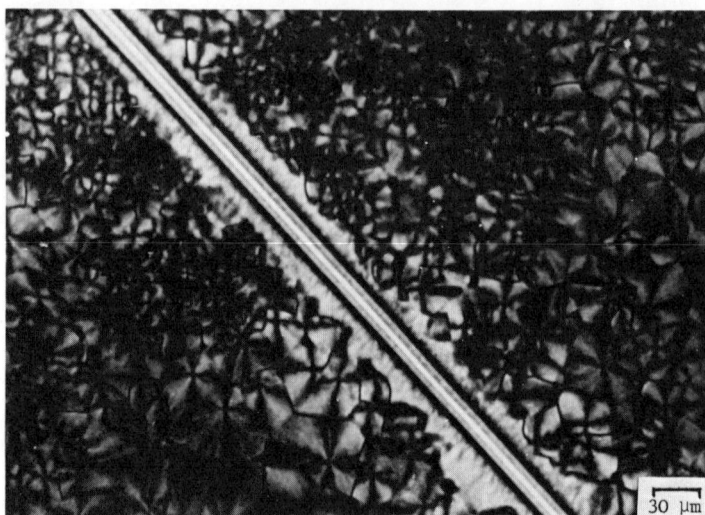

Fig. 9.5. Optical micrograph showing uniform transcrystallinity around a Kevlar fibre in nylon 66.[10]

that of normal spherulites. The thickness of the transcrystalline layer can vary considerably for different polymers and thermal conditions. Under ideal conditions it may even be up to ten times the fibre diameter, but a more typical figure is of the order of one fibre diameter. The fact that the transcrystalline layer is anisotropic and can be of considerable thickness relative to the fibre suggests that there are good reasons to expect the layer to be of importance in modifying the fibre–matrix interfacial bond strength. We shall return to this aspect later in the chapter.

Transcrystallinity, as well as crystallinity in general, can be studied conveniently by the use of a microscope hot-stage. The technique has been described several times in the literature, and some earlier work is reported in refs 2 and 16–18.

A small amount of polymer, placed on a glass microscope slide with some fibres, is heated on the hot-stage for 15–30 min at 30–40 °C above the melting point, to remove all existing crystallinity. Where necessary (e.g. with nylon), to minimise degradation of the polymer during this period, air is excluded by circling the polymer with a small amount of silicone fluid (e.g. MS 550), and placing a cover slip over it. As soon as the polymer melts, the cover slip is pressed down to seal out air and reduce the thickness of the polymer. After this pre-heating period, the slide is transferred rapidly to a second hot-stage maintained at the required crystallising temperature. Alternatively, the same hot-stage can be cooled down. When the crystallisation has reached the desired state, which is typically after a few minutes (but may vary from seconds to hours) as observed through the polarizing microscope, the specimen is quenched in iced water or, better still, in dry ice and acetone.

An alternative procedure for producing transcrystallinity around fibres was adopted by Bessel *et al.*[19, 20] who investigated transcrystallinity by the *in situ* polymerisation of a nylon 6 matrix around aligned glass and carbon fibres. There are, however, some disadvantages in this method in that the crystallisation temperature cannot be varied independently of the polymerisation temperature and there is also some doubt about the degree of polymerisation close to the fibres.[21] Bessel *et al.* identified two zones around the fibres: an inner zone with a fine random structure, and an outer zone with a fibrillar structure extending to the spherulitic region. The inner zone was attributed to rapid nucleation of the nylon on the carbon fibre surface, and is more pronounced on Type I fibres.

Crystallinity can be produced over a range of temperature, but even with careful control of conditions on the hot-stage there is often considerable variation in the amount of transcrystallinity along the fibres. Some of this variability may well be due to lack of uniformity of the fibre surface, caused by any number of factors, e.g. inconsistent surface treatment, sizing and possibly damage due to rubbing on guides during manufacture, it being known that the nucleating power of a surface is critically dependent on its topography. The present authors[10] have found, for example, that the strong nucleating power of Kevlar fibre as shown in Fig. 9.5 can be reduced very appreciably by pounding it manually with a pestle and mortar. Burton *et al.*[22] have also observed a further possible cause of variation. As we have seen above, nucleation and columnar growth is temperature dependent, and apart from the outer surface of the matrix polymer, the temperature is normally substantially uniform throughout the bulk. However, carbon fibres are highly conductive, and thus appreciable temperature gradients can exist close to their surfaces, well into the interior of the sample. The transcrystalline growth along a high-modulus (HM) carbon fibre, part of which is in contact with the outer regions of the sample can thus vary appreciably (Fig. 9.6). By way of comparison, Fig. 9.7 shows very

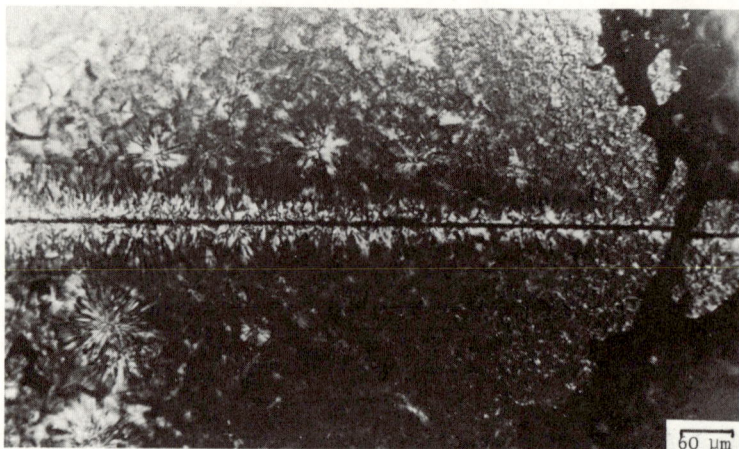

**Fig. 9.6.** Optical micrograph showing the variation in transcrystalline growth due to a temperature gradient along an HM carbon fibre in polypropylene.[22]

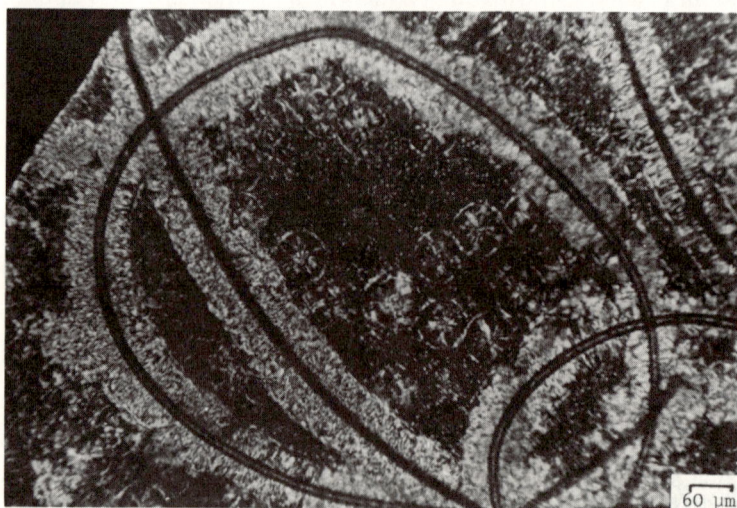

**Fig. 9.7.** Optical micrograph showing for comparison with Fig. 9.6 the very uniform transcrystalline growth along Terylene fibres.[22]

uniform transcrystallinity on Terylene fibres in which the thermal conductivity is much lower.

Folkes and Hardwick have also shown[23] that transcrystalline growth can be dependent on the molecular weight of the matrix polymer.

Polypropylene of different molecular weights showed considerable differences in the amount of growth. On the other hand it has been shown[24] that nylon exhibits vigorous transcrystallinity over a very wide range of molecular weights. Several authors have investigated epitaxy at surfaces as a major factor in the nucleation process. We have already mentioned the work of Turnbull and Vonnegut;[9] Hobbs[25] has also suggested, for example, that the large difference in nucleating ability between high-modulus (Type I) and high-strength (Type II) carbon fibres in polypropylene can be explained by differences in the size and orientation of their constituent graphite planes. The pronounced nucleation effect observed with Type 1 fibres was associated with the epitaxial growth of the matrix on to the large graphite planes present at the fibre surface. However, this again cannot provide a generally valid explanation since pronounced transcrystallinity is also observed with other polymers, such as nylon 66, where there is considerable mismatch in the lattice parameters of the matrix and substrate. Nevertheless, Hobbs has demonstrated very convincingly in a further paper[26] that the topography of the surface is almost certainly a major factor, by making a replica of a nucleating surface with non-nucleating material and then showing that this material becomes strongly nucleating.

## 9.4. MICROSTRUCTURE OF MOULDED COMPONENTS

Hot-stage microscopy, as described above, provides a useful method for detailed studies of the nucleating effects of fibre surfaces. The microstructure of components moulded in short fibre reinforced thermoplastics will necessarily be much more complex, with transcrystallinity possibly playing a minor role in some circumstances. As we shall see, a number of rather unexpected morphological features can arise as a result of more general material processing. For example, in the case of injection moulding, it is well known that very significant fibre breakage can occur, which will lead to a deterioration in the mechanical properties of the component. When the reinforcement is carbon fibre, it has been found that a significant volume of debris can be produced, arising from the fracture of the fibres.[10] Small angular particles are seen distributed throughout the matrix in addition to the fibres. A typical result for carbon fibre reinforced nylon 66 is shown in Fig. 9.8. These debris particles can act as a competing source of nucleation and are

Fig. 9.8. Optical micrograph of a thin section cut from a carbon fibre reinforced nylon 66 injection moulded component showing the presence of debris arising from fibre attrition.[10]

believed to be partly responsible for the progressive difficulties encountered in developing transcrystallinity around carbon fibres as the concentration of fibres in the composite increases. As a consequence of the high density of nucleation in the matrix, the spherulite dimensions are much smaller than those observed in virgin nylon 66. This observation also raises the question of the validity of using matrix data for the prediction of the mechanical properties of carbon fibre reinforced thermoplastics.

Irrespective of the fibre type being considered, it is well known that complex patterns of fibre orientation will develop in moulded components. In addition, however, it is known that relative movement of the fibres with respect to the melt occurs during processing. Hence, during the moulding process the fibres will create a localised perturbation of the overall flow field. One may anticipate that these microrheological effects could be very significant in influencing the morphology of the matrix during subsequent crystallisation. For example, they may directly affect the development of any transcrystalline layer around the fibre or be responsible for the formation of other morphological features which are fibre related.

Although these considerations may be undesirable complications in understanding the structure–property relationships in reinforced thermoplastics, they may well offer routes for the significant enhancement of mechanical properties in reinforced grades of polymers. At the very least, they demonstrate the need to study the interaction of the fibres with the matrix in a quite general way. Accordingly, we will now discuss the general flow features, e.g. fibre orientation observed in injection moulded bars, and then separately discuss localised flow phenomena around fibres, and their consequences.

### 9.4.1. General Flow Features in Fibre Reinforced Thermoplastics

We are concerned here with the fibre orientation distribution developed in an injection moulded component, and the relationship, if any, to the overall pattern of molecular orientation as assessed using polarising microscopy. As a simple example, we will consider the fibre orientation in a simple parallel-sided bar which is gated at one end: see Fig. 9.9. To avoid effects due to fibre debris formation, such as occur when carbon fibres are the reinforcement, we choose for our initial description of orientation the case of glass fibre reinforced polypropylene. The fibre orientation distribution in such a bar has been discussed by Bright *et al.*[27] and Folkes and Russell.[28] For high injection speeds such as those

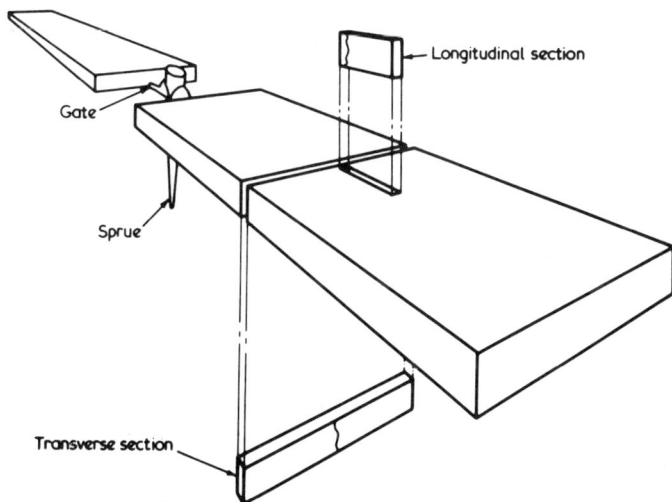

**Fig. 9.9.** Two-cavity strip moulding showing locations of the transverse and longitudinal sections used for birefringence and fibre orientation studies.[28]

used in the commercial moulding of fibre reinforced thermoplastics, the
fibre orientation pattern in the bar takes the form as shown in Fig. 9.10.
A layered structure is observed; in the core of the moulding the fibres
are mainly aligned perpendicular to the flow direction. Above and
below this are regions with the predominant fibre orientation in the flow
direction. This skin–core microstructure is a feature of all injection
moulded artefacts although the precise pattern of fibre orientation in
the skin and core will vary according to the material and processing
conditions.

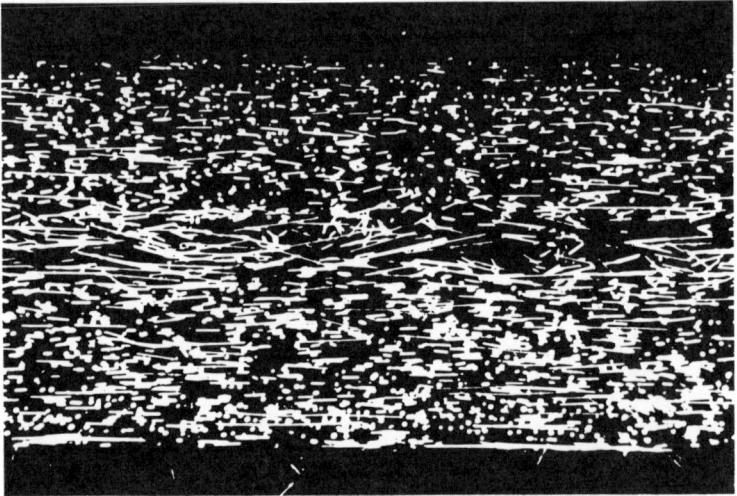

**Fig. 9.10.** Contact microradiograph showing the skin–core microstructure in a
transverse section cut from the strip moulding for fast injection.[27]

The same sections used above for studying fibre orientation have also
been examined in transmitted polarised light.[28] Figure 9.11 shows the
variation of birefringence, $\Delta n$, across both the transverse and longitudinal
sections cut from the moulded bar. It can be seen that the birefringence
is very small for the transverse section but varies in a pronounced
manner across the longitudinal section. It shows that the molecular
orientation is high close to the surfaces of the moulding, and is
predominantly along the bar axis. This molecular orientation is
attributed to the elongational flow occurring at the advancing melt front
during mould filling. The subsidiary maxima in $\Delta n$ away from the

surface are attributed to molecular orientation induced by shear flow close to the boundary between the solidified skin and molten core material.[29] It is interesting that these peaks occur at a position where

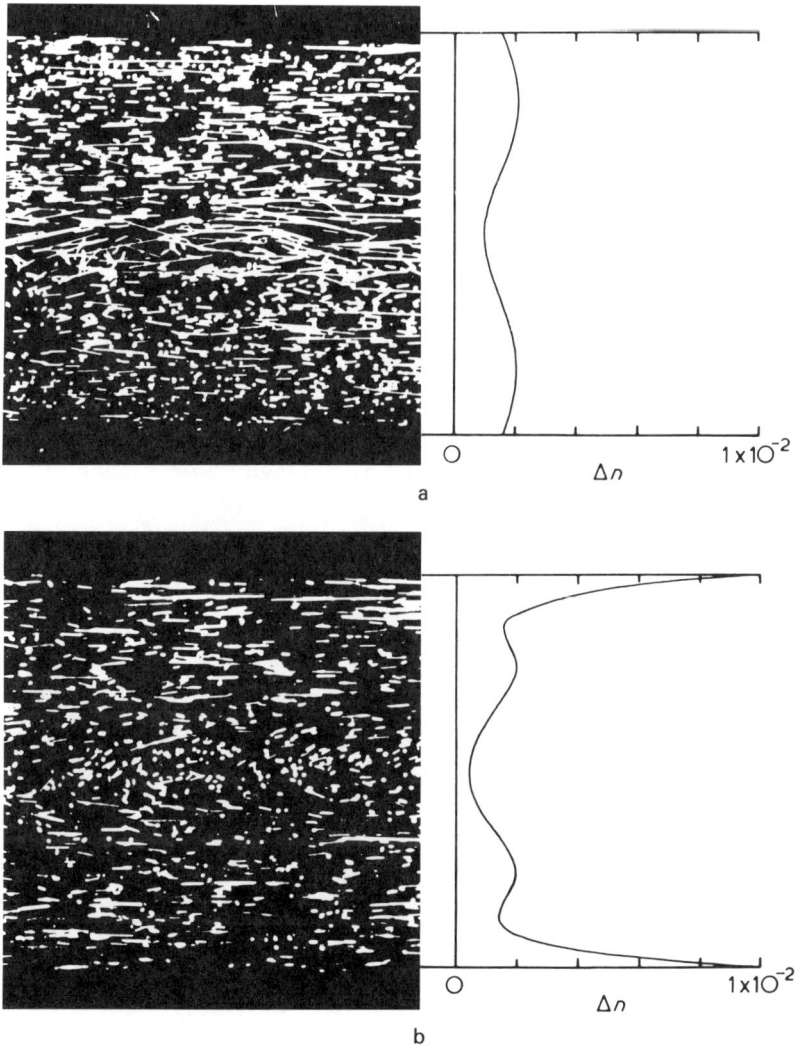

Fig. 9.11. Birefringence variation through the thickness of the strip moulding and associated contact microradiographs for fast injection: (a) transverse section; (b) longitudinal section.[28]

there is no real obvious change in fibre orientation, as judged from the micrograph in Fig. 9.10.

Of key overall importance, however, is the fact that birefringence values of magnitude comparable with those in unfilled melts are observed in this material and other fibre reinforced thermoplastics. The implications of these findings for the interpretation of the anisotropy of the physical properties in mouldings is self-evident, and indicates in particular that a formal analysis of the mechanical anisotropy will require composite theories which allow for the matrix phase being anisotropic. Detailed comparisons of birefringence and fibre orientation data of the type described indicate that for the range of fibre concentrations normally encountered in commercially available grades of thermoplastics, there is little direct overall relationship between fibre orientation and molecular orientation. This suggests that although the fibres will modify the gross rheological properties of the matrix phase they have little effect on the molecular orientation in the matrix.

As the fibre concentration in the composite is increased, the situation described above changes considerably. As an example, Fig. 9.12 shows the variation of birefringence across a longitudinal section of a moulding of short glass fibre reinforced polyethylene for different fibre concentrations.[28] The subsidiary peaks in $\Delta n$, associated with the

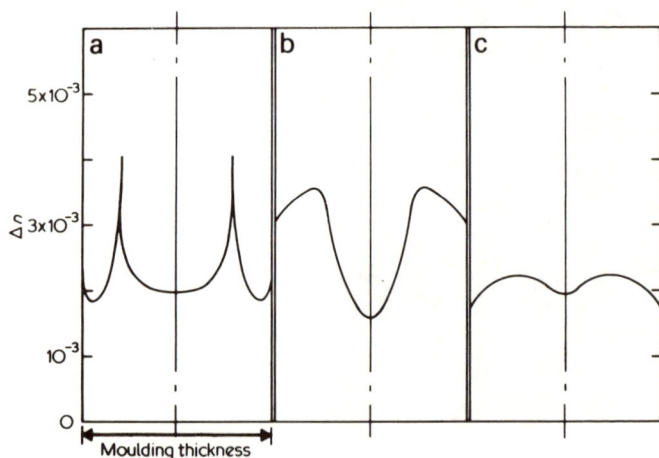

Fig. 9.12. Birefringence variation through the thickness of the strip moulding: (a) unfilled polyethylene; (b) polyethylene containing 20 wt % of glass fibres; (c) polyethylene containing 50 wt % of glass fibres.[28]

shearing of the melt at the skin–core boundary, become less distinct and broader as the fibre fraction increases. As we shall see, this is contrary to what is expected from a knowledge of the rheological properties of polymer melts containing high concentrations of fibres. Thus, as the fibre concentration increases, the velocity profile of the melt, measured under isothermal conditions, changes from a pseudo-parabolic profile for zero fibre concentration to almost complete plug flow at high concentrations.[30] With reference to Fig. 9.13, showing a schematic diagram of the mould filling process during injection moulding, it can be seen that the molten core of material flows within a solid skin layer. The detailed shape of the velocity profile for the core depends on the rheological characteristics of the fluid, the flow rate and the temperature of the incoming melt, as well as the temperature of the mould cavity. To a first approximation, at least, this part of the flow process is frequently regarded as capillary flow. Accordingly, one would expect that as the fibre concentration increases, the region between the skin and core over which an appreciable shear rate exists would become narrower, and hence the associated peaks in $\Delta n$ should also become much narrower. This result is not in accord with the experimental data. This discrepancy implies that it is incorrect to interpret the variation of $\Delta n$ in terms of the macroscopic flow properties of the melt. In fact, qualitatively, the pattern of variation of $\Delta n$ can be explained by assuming that it is due entirely to columnar spherulitic growth of the matrix around the fibres, as would be observed using hot-stage microscopy. Thus, in this case, the variation in $\Delta n$ in the moulding would be directly related to the degree of fibre orientation along the principal flow direction. Therefore, the maximum value of $\Delta n$ obtainable would correspond to fully aligned fibres, while fibres aligned at $90°$ to the principal flow direction would result in optical isotropy for longitudinal sections. An examination of

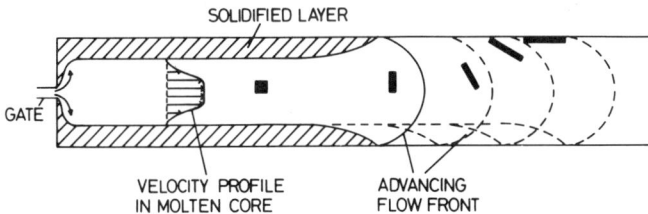

Fig. 9.13. Schematic diagram of the mould filling process showing the deformation of an initially square fluid element at successive positions of the advancing flow front.[28]

the microstructure of the bar used for these studies confirmed that the pattern of fibre orientation and birefringence were indeed very similar for the highest concentration of fibres studied.[28]

At the time the above work was carried out, no detailed study of the microstructure of the matrix around individual glass fibres was carried out, partly because of the practical difficulty of producing the very thin sections needed for such an investigation. Recently, however, Misra et al.[31] have reported the direct observation of transcrystallinity in injection moulded glass fibre reinforced polypropylene. They found that as the fibre volume fraction was increased, transcrystallinity developed around almost all the fibres. This finding is consistent with the results of Folkes and Russell,[28] although the matrix material was different in the two cases.

In fact, it transpires that transcrystallinity in moulded artefacts is not the only morphological feature to be of relevance as far as the general subject of fibre–matrix interactions is concerned. Indeed, there can be other features occurring even more readily in some cases, but which none-the-less are directly related to the presence of the fibres.

### 9.4.2. Localised Melt Flow Phenomena in Fibre Reinforced Thermoplastics

The growth morphology and rate of crystallisation of polymer molecules from both solutions and melts are greatly affected by the presence of a hydrodynamic field. Deformation and orientation of the molecules, brought about by the action of flow-induced stresses, can lead to nucleation and growth of extended-chain crystalline structures at elevated temperatures, whereas predominantly chain-folded lamellar or spherulitic structures form under quiescent conditions. The shish-kebab morphology characteristic of flow-induced growth is also associated with the high tensile modulus and strength and other favourable mechanical properties of highly oriented systems.[32] A key feature of the formation of fibrous crystals is that the nucleation process is related to the presence of regions of extensional flow, while continued growth of an already nucleated or seeded fibre can occur in predominantly shearing flow.[33] A fibre or other slender body immersed in a melt will create its own local flow field, which is considerably different from the macroscopic field. Mackley[34] has analysed the flow past a fibre, modelled as a prolate spheroid, and has shown that along the symmetry axis of the spheroid (fibre axis) an extensional flow field can be generated with extension rates as high as $10^4 \, V \, \mathrm{s}^{-1}$ near the tip, where $V$

is the overall fluid velocity parallel to the symmetry axis. Over the side surface of the fibre, shearing flow exists with shear rates at least an order of magnitude smaller than the extension rates at the tip. More recently, Dairanieh and McHugh[35] have extended the analysis to take into account the effects of different external flow kinematics on the local flow. The importance of extensional flow in producing chain alignment has been demonstrated experimentally by Mackley *et al.,*[36] and this type of flow field is also responsible for the successful production of ultra-high-modulus polyethylene fibres.[32]

In the context of short fibre reinforced thermoplastics it is clear that the above considerations could be highly relevant, not only in providing insight into crystallisation phenomena around fibres in moulded artefacts but as a possible route for enhancing the mechanical properties of the composite. In fact, the consequences of extensional flow, produced at the fibre ends, has been recently observed in at least two composite systems; HM carbon fibres in polypropylene[22] and Terylene fibres in polypropylene. Figure 9.14 shows an optical micrograph taken in polarised light of a thin section cut from an injection moulded ASTM bar of HM carbon fibre reinforced polypropylene. Shish-kebabs can be observed originating from specific fibres in the section. They can

Fig. 9.14. Optical micrograph showing the presence of shish-kebabs in an injection moulded bar of HM carbon fibre reinforced polypropylene.[22]

be seen in even greater profusion when Terylene rather than carbon fibres are used as reinforcement. A diagram showing the mode of formation of the shish-kebab is shown in Fig. 9.15. In the former case, detailed measurements of the optical properties of the backbone of the shish-kebab have been made.[22] In particular, the retardation was found to vary over a very large range, 12–470 nm. The width of the backbone was found to be ~0·4 $\mu$m. If this backbone is assumed to be of circular cross-section, then even the lowest measured retardation gives a birefringence value corresponding closely to fully oriented polypropylene chains, i.e. $\Delta n = 30 \times 10^{-3}$. In order to account for the very large retardation values observed, it is proposed that those fibres inclined to the surface of the section have produced sheets of chain-extended material, as shown schematically in Fig. 9.16. The maximum depth of the sheets will of course be equal to the section thickness, which

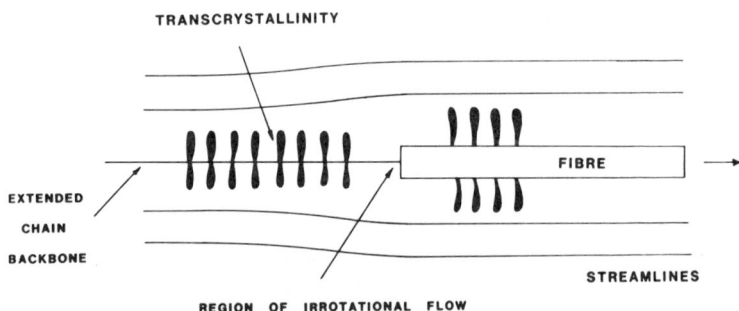

Fig. 9.15. Schematic diagram showing the proposed mechanism of formation of shish-kebabs in fibre reinforced thermoplastics.[22]

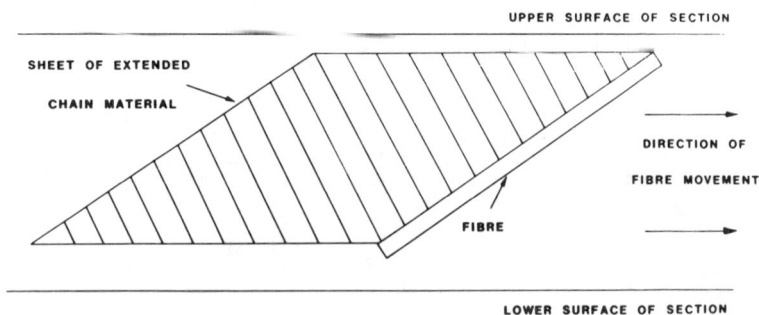

Fig. 9.16. Schematic diagram showing the proposed mechanism of formation of sheets of chain-extended material in fibre reinforced thermoplastics.[22]

in the case of the above investigation was 15 μm. It then follows that the maximum retardation value is associated with a path length of not 0·4 μm but 15 μm, thereby giving a birefringence value within the accepted range for extended-chain polypropylene.

The conditions leading to the formation of chain-extended entities in fibre reinforced thermoplastics have not yet been systematically studied However, considering the very large concentrations of fibres used in practice, it is likely that the effects described above could be of considerable practical importance. In addition to the obvious relevance of extensional flow in fibre reinforced thermoplastics, there are indications that the shearing flow occurring at the fibre–matrix interface during moulding could also be of importance in controlling and modifying the local morphology around a fibre. It is not easy to anticipate the effects that will take place since the extent of any transcrystallinity will be influenced by both the fibres and the nucleation processes in the matrix as a whole. Lagasse and Maxwell[37] have shown that the rate of nucleation in a polymer is increased by flow, and this would imply that transcrystalline growth could be impeded by the competing spherulitic growth in the matrix. On the other hand, Misra *et al.*[31] have reported results to the contrary. They studied the interfacial microstructure around glass fibres in polypropylene, using hot-stage microscopy. It was found that transcrystalline regions developed only around those fibres which were deliberately pulled through the matrix prior to crystallisation, whilst no crystallisation occurred along the undisturbed fibres. It would seem that the ability of the fibres to nucleate the matrix is promoted by the development of a shear stress at the interface. This is in agreement with the observations made by Gray.[38]

The promotion of fibre nucleation by melt flow, as described, may be related to some rheological data reported by Burton *et al.*[39] Measurements of the rheological properties of polystyrene of various molecular weights were conducted with a Weissenberg rheogoniometer using extremely small parallel plate separations (20–500 μm). The data indicated that the viscosity of the melt became appreciably inhomogeneous across the gap between the plates as the molecular weight of the polymer was increased. On a molecular basis, one model consistent with this result is shown schematically in Fig. 9.17. This experiment partially mimics the situation that could arise between two adjacent fibres in a fibre reinforced thermoplastic undergoing melt flow. The model allows for the possibility that, during flow, those molecules

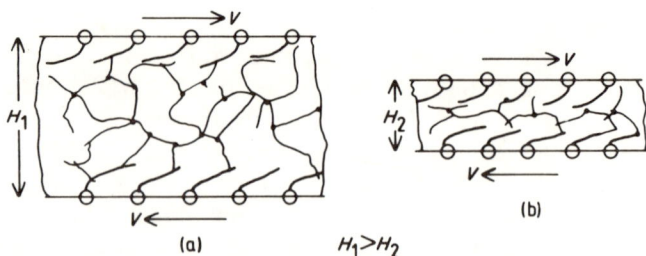

**Fig. 9.17.** Schematic diagram illustrating one class of conformation of polymer molecules in the gap between the parallel plates of the Weissenberg rheogoniometer producing lower viscosity near the surfaces: (a) wide gap; (b) narrow gap.[39]

already adsorbed on the rotor (fibre) will deform, thereby generating new attachments on the rotor (fibre). In this way the number of potential nucleation sites can grow as flow proceeds. Although this work was conducted using a non-crystalline polymer, there is every reason to expect that the proposed model would also be appropriate for semi-crystalline polymers.

## 9.5. THE EFFECT OF INTERFACIAL MICROSTRUCTURE ON COMPOSITE PROPERTIES

Since the mechanical properties of a composite material depend critically on the nature of the interface, it follows that there are good reasons to expect any special microstructural features to be important. For example, when transcrystallinity occurs at a fibre surface, the fibre will be surrounded by a sheath of polymer, whose properties will differ from those of the remaining matrix material. Furthermore, this layer will be anisotropic due to the molecular orientation. Considering that, in many instances, the thickness of the transcrystalline layer is at least equal to one fibre diameter, one can expect that in many composites of commercial importance the layers around adjacent fibres will impinge. In this case, not only would the fibre–matrix interface be modified but also the entire matrix.

There have been a number of rather disparate observations concerned with the relationship between interfacial microstructure and properties but, overall, experimental data in this area are very limited. One of the earliest systematic studies was reported by Bessel *et al.*[20] Using nylon 6 as the matrix, they prepared aligned continuous fibre composites. Both glass and carbon fibres were used, up to a volume fraction of 15%. The glass fibres were found to be comparatively ineffective in promoting transcrystallinity compared with the carbon fibres. They found good agreement between the experimental values of longitudinal tensile modulus and strength and those values predicted using the simple rule of mixtures. Observation of the fracture surfaces of the glass fibre composites showed that very extensive fibre pullout had occurred with the fibres surrounded by sheaths of nylon. It was also noted that, in these specimens, the propagating cracks were occasionally redirected parallel to the fibres at a distance of about 2 $\mu$m from the actual fibre-matrix interface. These workers suggested that there is a surface of weakness some distance from the fibre surface which may be related to the columnar region around the fibre.

Campbell and Qayyum[40] found that the incorporation of nylon and Terylene fibres into polypropylene led to the expected increase in tensile strength according to the rule of mixtures. Elongation to break, however, was significantly greater, and increased with volume fraction of fibres. The authors attributed this to the decrease in spherulite size of the polypropylene, and to a restraint on the necking of the fibres due to transcrystalline growth at the fibre surface.

Ritchie and Cherry[41] and Masuoka[42] have also investigated the possible role of transcrystallinity as a factor in controlling adhesive bond strengths. In the former case, lap shear tests were performed using glass blocks as substrates and high-density polyethylene as the adhesive. The joint formation conditions were varied in an attempt to provide samples having a range of interfacial microstructures. A large scatter in the values of the lap shear strengths was found, from which they conclude that the presence (or otherwise) of transcrystallinity does nothing to improve bond strength. Masuoka[42] performed tensile bond strength measurements on mild steel substrates bonded using nylon 12. As with Ritchie and Cherry,[41] the interfacial morphology and degree of crystallinity of the nylon 12 were varied by changing the cooling rate from the melt during sample preparation. It was concluded that changes in bond strength arose from the change in degree of crystallinity rather than from the transcrystallinity *per se*. In spite of these observations, however, the potential relevance of transcrystallinity in fibre reinforced thermoplastics remains. For example, Kardos[43] has examined the microstructure and mechanical properties of glass and carbon fibre reinforced polycarbonate. Using a special moulding and annealing sequence, he has reported that the tensile modulus and strength of test bars can nearly double compared with those values obtained using standard moulding conditions. With some supportive evidence, he concludes that this enhancement of properties is due to the localized formation of crystalline entities on the fibre surface. In fact, the argument draws heavily on the still controversial issue of whether small regions of crystalline order exist in those polymers normally regarded as amorphous.

## 9.6. CONCLUSIONS

In this chapter we have seen that the introduction of reinforcing fibres can affect the surrounding polymer matrix in several ways on the molecular scale. With semicrystalline polymers, columnar growth may occur on the fibres, thus possibly influencing the bond strength, the strength at the sheath–bulk polymer interface, and the general reinforcing effect of the fibre. The crystallinity of the bulk of the matrix may also be influenced by the presence of fibres.

The localised melt flow occurring around the fibres during the moulding of the reinforced thermoplastic is important for the formation of chain-extended entities in the matrix and in possibly influencing the nucleation processes involved in transcrystalline growth.

It may be possible, by varying the thermal and mechanical conditions empirically, to arrive at optimum conditions for the production of fibre reinforced thermoplastic components. However, in the long term, it is almost always preferable to achieve an understanding of all the factors and phenomena involved on a rational scientific basis, so that lasting improvements in material formulation and processing technology can be obtained.

## REFERENCES

1. Hartshorn, N. H. and Stuart, A., *Crystals and the Polarising Microscope,* 4th edn, Edward Arnold, London, 1970, p. 577.
2. Cannon, C. G., Chappel, F. P. and Tidmarsh, J. I., *J. Textile Inst.,* 1963, **54**, T210.
3. Mann, J. and Roldan-Gonzalez, L., *J. Polym. Sci.,* 1961, **60**, 1.
4. Khoury, F., *J. Polym. Sci.,* 1958, **33**, 389.

5. Keith, K. D. and Padden, F. J., *J. Polym. Sci.*, 1959, **39**, 101.
6. Chatterjee, A. M., Price, F. P. and Newman, S., *J. Polym. Sci., Polymer Phys. Ed.*, 1975, **13**, 2369.
7. Campbell, D. and Qayyum, M. M., *J. Polym. Sci., Polym. Phys. Ed.*, 1980, **18**, 83.
8. Way, J. L., Atkinson, J. R. and Nutting, J., *J. Mater. Sci.*, 1974, **9**, 293.
9. Turnbull, D. and Vonnegut, B., *Ind. Eng. Chem.*, 1952, **44**, 1292.
10. Burton, R. H. and Folkes, M. J., *Plast. Rubber Process. Appl.*, 1983, **3**, 129.
11. Pruneda, C. D., Steele, W. J., Kershaw, R. P. and Morgan, R. J., *Composites Techn. Review*, 1981, **3**, 103.
12. Kontsky, J. A., Walton, A. G. and Baer, E., *J. Polym. Sci. Part B*, 1967, **5**, 185.
13. Binsbergen, F. L., *J. Polym. Sci., Polymer Phys. Ed.*, 1973, **11**, 117.
14. Keller, A., *Reports on Progress in Physics*, 1968, **31**(2), 623.
15. Keller, A., *J. Polym. Sci.*, 1955, **15**, 31.
16. Magill, J. H., *Polymer*, 1961, **2**, 221.
17. Khoury, F., *J. Polym. Sci.*, 1957, **26**, 375.
18. Keller, A., *J. Polym. Sci.*, 1955, **17**, 291.
19. Bessel, T. and Shortall, J. B., *J. Mater. Sci.*, 1975, **10**, 2035.
20. Bessel, T., Hull, D. and Shortall, J. B., *Faraday Spec. Discuss. Chem. Soc.*, 1972, **2**, 137.
21. Frayer, P. D. and Lando, J. B., *J. Coll. Interface Sci.*, 1969, **31**, 145.
22. Burton, R. H., Day, T. M. and Folkes, M. J., *Polymer Communications*, 1984, **25**, 361.
23. Folkes, M. J. and Hardwick, S. J., *J. Mater. Sci. Letters.*, 1984, **3**, 1071.
24. Burton, R. H. and Folkes, M. J., unpublished work.
25. Hobbs, S. Y., *Nature, Phys. Sci.*, 1971, **234**, 12.
26. Hobbs, S. Y., *Nature, Phys. Sci.*, 1972, **239**, 28.
27. Bright, P. F., Crowson, R. J. and Folkes, M. J., *J. Mater. Sci.*, 1978, **13**, 2497.
28. Folkes, M. J. and Russell, D. A. M., *Polymer*, 1980, **21**, 1252.
29. Tadmor, Z., *J. Appl. Polym. Sci.*, 1974, **18**, 1753.
30. Crowson, R. J., Folkes, M. J. and Bright, P. F., *Polym. Eng. Sci.*, 1980, **20**, 925.
31. Misra, A., Deopura, B. L., Xavier, S. F., Hartley, F. D. and Peters, R. H., *Angew. Macromol. Chem.*, 1983, **113**, 113.
32. Ciferri, A. and Ward, I. M. (Eds), *Ultra-High Modulus Polymers*, Applied Science Publishers, London, 1979.
33. McHugh, A. J., *Polym. Eng. Sci.*, 1982, **22**, 15.
34. Mackley, M. R., *Colloid Polym. Sci.*, 1975, **253**, 373.
35. Dairanieh, I. S. and McHugh, A. J., *J. Polym. Sci., Polym. Phys. Ed.*, 1983, **21**, 1473.
36. Mackley, M. R., Frank, F. C. and Keller, A., *J. Mater. Sci.*, 1975, **10**, 150.
37. Lagasse, R. R. and Maxwell, B., *Polym. Eng. Sci.*, 1976, **16**, 189.
38. Gray, D. G., *J. Polym. Sci., Polym. Letters Ed.*, 1974, **12**, 645.
39. Burton, R. H., Folkes, M. J., Narh, K. A. and Keller, A., *J. Mater. Sci.*, 1983, **18**, 315.
40. Campbell, D. and Qayyum, M. M., *J. Mater. Sci.*, 1977, **12**, 2427.

41. Ritchie, P. J. A. and Cherry, B. W., 'Transcrystallinity as a factor in controlling adhesion', *13th and 14th Conferences on Adhesion (City University, London) 1975/76,* Ch. 15, p. 235.
42. Masuoka, M., *Int. J. Adhesion and Adhesives,* 1981, **1,** 256.
43. Kardos, J. L., *J. Adhesion,* 1973, **5,** 119.
44. Pennings, A. J., Van der Mark, J. M. A. A. and Kiel, A. M., *Kolloid Z.u.Z. Polymere,* 1970, **37,** 336.

*Chapter 10*

# Reinforced Thermoplastic Foams

G. C. McGrath and D. W. Clegg

*Department of Metals and Materials Engineering, Sheffield City Polytechnic, Sheffield, UK*

## 10.1. INTRODUCTION

For many years glass fibre reinforced thermoplastics have successfully bridged the materials gap between unreinforced thermoplastics and the more conventional engineering materials such as steel and wood. More recently structural foams have been developed to reduce the weight of plastic products, at the same time providing a one-step route to a twin-component structure which could maximise the properties per unit weight at a minimum cost.

An interesting development of these trends has been to consider combination of the two concepts (fibre glass reinforcement and foaming) and to use, for example, short glass fibres to minimise or even reverse some of the property sacrifices incurred in structural foaming.[1] A comparison of the results of this concept is shown in Table 1.

The reinforced structural foam has mechanical properties superior to those of the rigid resin, but the high percentage of voids when glass fibres are present in the foamed high density polyethylene results in a density comparable to that of the unreinforced foam, giving a higher strength-to-weight ratio. It must be emphasised, however, that to achieve satisfactory properties, the reinforcement needs treating with a suitable coupling agent for the thermoplastic being considered (Fig. 10.1).

## 10.2. FABRICATION METHODS

Injection moulding is used almost exclusively for fabrication although extrusion is possible. Thermoplastic foams are prepared by introducing a gas into the polymer melt and allowing the gas to expand, so displacing the thermoplastic resin. Close control of the process is essential to form a fine, uniform cell structure as there is always the possibility that this structure will collapse to form one large void.

Many foaming processes have been developed and are available commercially. Most of the processes are suitable for use with glass fibre reinforced thermoplastics providing the equipment manufacturer is advised of the intention to use glass fibres, so that the moulds and machines are built with steels suitably treated to prevent excessive abrasion[2] although this largely depends upon the thermoplastic being processed. Polypropylene–glass fibre compounds are virtually non-abrasive, but when nylon–glass compounds are processed abrasion can be a problem because of the increased viscosity. These are the two extremities of the scale, with other thermoplastics falling somewhere between.

These processes involve either low or medium-to-high pressure systems. Typical pressures involved are approximately one-tenth and one-third, respectively, of the solid injection moulding pressures.

TABLE 1
Comparison of the Properties of High-density Polyethylene (HDPE), Glass Fibre Reinforced HDPE and Their Foamed Versions

| Property | ASTM method | Units | Solid HDPE | Solid HDPE +15% glass fibre | Foamed HDPE | Foamed HDPE +15% glass fibre |
|---|---|---|---|---|---|---|
| Tensile strength | D638 | MPa | 18·7 | 41·9 | 16·7 | 23·0 |
| Tensile modulus | D638 | GPa | — | 3·35 | 0·97 | 1·74 |
| Elongation | D638 | % | — | 6·7 | 8·8 | 5·8 |
| Flexural strength | D790 | MPa | 18·1 | 54·3 | 18·1 | 30·9 |
| Flexural modulus | D790 | GPa | 0·8 | 2·54 | 0·82 | 1·54 |
| Solids | D2584 | mass % | 0 | 15·9 | 0 | 14·8 |
| Density | D792 | kg m$^{-3}$ | 960 | 1·070 | 880 | 900 |
| Calculated density reduction | — | vol. % | 0 | 0 | 10 | 18 |

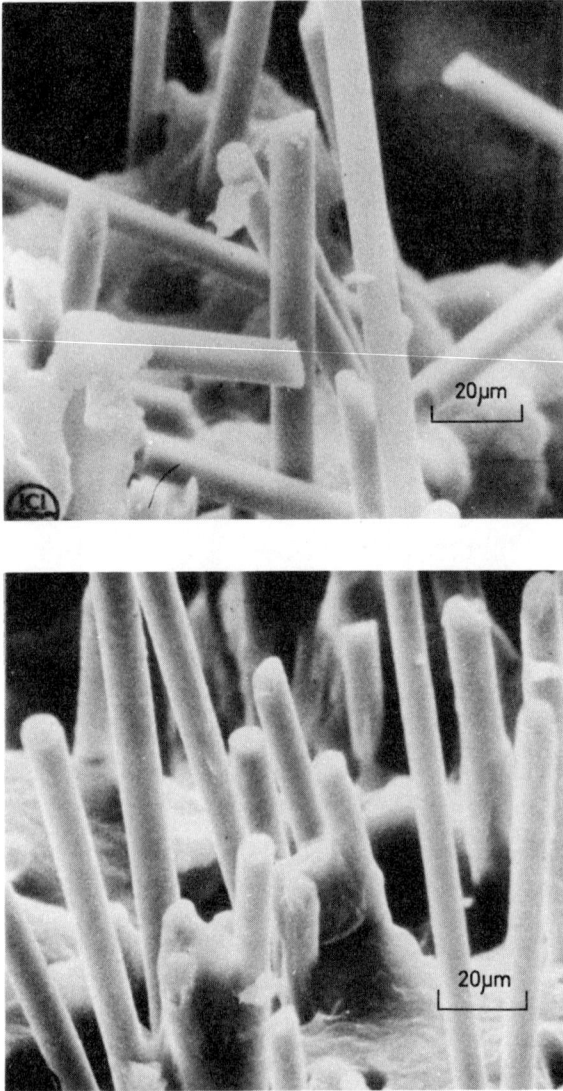

Fig. 10.1. Scanning electron micrographs of fractured specimens of injection moulded, reinforced polypropylene foam. *Above*: without coupling agent — during fracture, some of the fibres have slipped out of engagement with the matrix. *Below*: with coupling agent — the retention of fibres in the polymer after fracture demonstrates that bonding between fibres and matrix was well established. (Courtesy Imperial Chemical Industries PLC.)

## 10.2.1. Low Pressure Processes

These are used mainly in Europe: the basic principle is that a quantity of melt, containing a dissolved inert gas, less than that required to fill the mould is injected so that expansion proceeds under lowered pressure and the fixed mould is filled. As the dissolved gases come out of solution, foaming results. A solid skin is formed on the outside surface as the cells collapse in this region.

Three variations of the basic process dominate current practice. They are as follows.

(i) Solid chemical blowing agent is mixed with polymer and processed in specially constructed moulding machines. The main features of these machines are an efficient shut-off nozzle system and low mould locking forces. The process is outlined in Fig. 10.2. In this instance, the gas is generated by the chemical decomposition of the blowing agent in the hot thermoplastic melt.

(ii) Solid chemical blowing agent is mixed with polymer and processed in a conventional high pressure injection moulding machine; critical modifications are an efficient shut-off nozzle and good control over back pressure. Because of the high locking forces, which are unnecessary for foams, capital and running costs for producing reinforced thermoplastic foams are higher than if a specially designed machine is used. Against this is the versatility offered by a choice of thermoplastic output, i.e. solid or foamed components.

(iii) The Union Carbide process: an extruder which usually has injection of an inert gas midway along the barrel runs continually to feed the pressurised melt to an accumulator, where it is held under pressure to keep the gas in solution until the charge volume is reached, whereupon it is injected into the mould which it proceeds to fill by expansion.

Structural foams have an extended cooling period which may affect adversely the economics of the proccess. One way of overcoming this is to use a multi-station rotary machine, which is particularly efficient for production of smaller parts.

## 10.2.2. Medium-to-high Pressure Processes

These processes are used where a better surface finish, comparable to solid mouldings, is required. The mould cavity is just filled, without

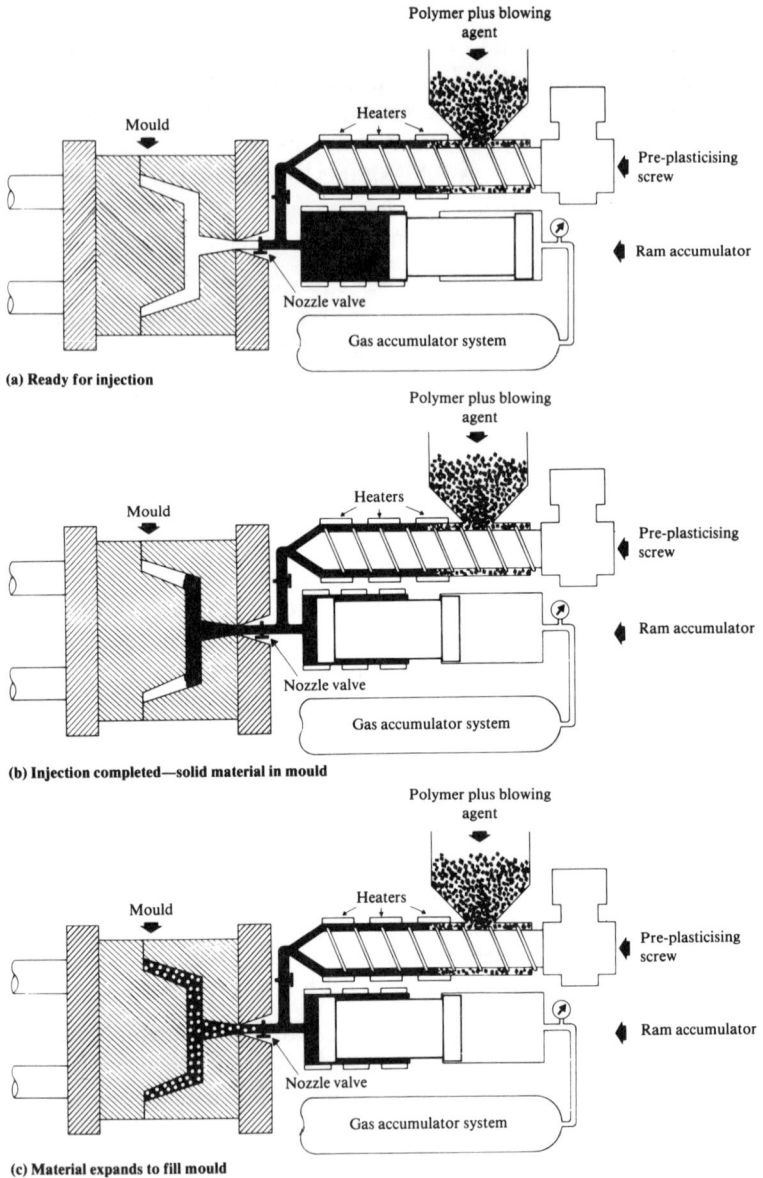

**Fig. 10.2.** Special purpose structural foam injection moulding machine for use with chemical blowing agents. (a) Ready for injection. (b) Injection completed; solid material in mould. (c) Material expands to fill mould. (Courtesy Imperial Chemical Industries PLC.)

allowing any foaming to take place. When full, the mould is expanded by withdrawing a segment or face of the mould in a controlled way which allows the foaming action to take place. The higher pressures used normally mean more robust and hence expensive moulds.

Numerous other processes have been developed in order to improve surface finish, including the following.

(i) The counter-pressure process. The cavity is pressurised by an inert gas prior to filling and so degassing is avoided. As injection proceeds a manifold is opened, thus allowing expansion of the thermoplastic melt and foam formation. The component densities are generally increased because of the increased skin thickness provided.

(ii) Thermal cycling of the mould, developed by BASF and Krauss-Maffei and known as the Vanthem process. Formation of a solid skin is prevented by heating the mould during injection. The skin is then formed by rapid cooling of the mould after injection of the thermoplastic melt.

## 10.3. MORPHOLOGY

As we have seen, thermoplastics are foamed by the introduction of gas into a melt. Reinforced thermoplastic foams are characterised by a density gradient from the core of the foamed structure to the surface. The cooling conditions produce this density gradient and variation of cooling rate produces different characteristics in the finished moulding.

Slow cooling allows greater time for cell growth and therefore the largest cells are found in the cooled region where heat dispersion has been slowest. On the contrary the cooler mould–melt interface promotes more rapid cooling rates, and this, together with the high shear rate along the mould walls, gives a solid skin of high density. Thus a density gradient is created. As the cells become smaller towards the surface of the moulding, the proportion of thermoplastic in the melt, and therefore its density, increases. Thus the physical properties of a foamed structure depend on the internal density gradient as well as the overall density.

It is essential to keep the cell formation under strict control at all times, to achieve a moulding with a high-density skin and a low-density core without large voids. The presence of glass fibres greatly aids this control. As the cells grow, the fibres are pushed back, mostly re-orientating themselves parallel to the cell walls. The orientation of fibres in the moulding is generally less ordered and more even than that found in solid injection mouldings. This produces more uniform properties in the finished component. A growing bubble uses part of its energy to push back the fibres and therefore will grow less than it otherwise would in an unreinforced foam and a fine cellular structure, free from voids produced by cellular collapse, is promoted. Glass fibres also increase the apparent viscosity and thermal conductivity of the melt. A combination of these effects leads to a better cell size control and more uniform physical properties.

Although properties are more uniform for a constant cross-section moulding, where section changes are encountered, it is usual for the thicker cross-sections to have somewhat lower densities than thinner cross-sections or areas near gates.[4] Parameters which have the greatest effect on this variation in density are the injection speed and final required weight of the moulding. Failure under tension in mouldings containing variations in cell density will nearly always be at the point of lowest density assuming constant cross-sectional area and that failure does not occur at a void.

We have described the density change from core to surface, but this is not the complete picture as the actual shapes of the cells will also vary being near-spherical in the core and elongated nearer the surface with virtually no cells at all in the skin region. However, careful examination of the cell structure in the skin region frequently shows a layer of extremely fine cells just behind the identifiable skin. These bubbles may be used as 'visual markers' indicating flow direction and flow of the wavefront during filling of the mould.[5]

## 10.4. INTERACTIONS BETWEEN PROCESSING AND MICROSTRUCTURES

To understand the advantages and disadvantages of glass reinforced thermoplastic foams it is instructive to consider how the cellular structure is formed during the moulding process. The saturation solution of thermoplastic, glass fibre and gas is allowed to expand in the mould. While the expansion is taking place the melt is cooling and solidifying and this stabilises the cell structure. The final structure will have a cell size distribution as seen in Fig. 10.3, with the cells at their

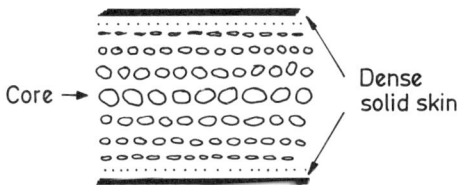

Fig. 10.3. Glass fibre reinforced thermoplastic foams showing cell size distribution through the moulding.

largest at the centre, decreasing in size towards the walls, and finally disappearing in the skin.

The reinforcing glass fibres have two significant effects on the mechanism of growth as follows.

(i) The apparent viscosity of the melt is increased and thus maintenance of a saturated melt solution is assisted since more energy is required for cell growth; this creates a fine cell structure and avoids cell collapse as the expansion proceeds.

(ii) A growing cell uses part of its energy to displace and partially orient the glass fibres and therefore the cell will grow less than it otherwise could in an unreinforced foam, creating a finer cellular structure, free of central voids. As the expansion proceeds the glass fibres are distributed in a partially random manner orientating themselves mostly tangentially to the cell walls as seen in Fig. 10.4. This produces cell walls reinforced with glass fibres and thus increased compressive strength. A high strength-to-weight ratio is produced in the composite.

## 10.5. MODEL SYSTEMS OF REINFORCED THERMOPLASTIC FOAMS

Glass fibre reinforced thermoplastic foams may be regarded as three-phase composites consisting of the following:

(i) a thermoplastic matrix;
(ii) glass fibre reinforcements; and
(iii) a cellular structure.

Thermoplastics and their reinforcements are discussed in Chapter 2, but the cellular structure is more difficult to define. However, the cellular portion of the composite can be considered simply as a filler and thus analysis of the composite is possible.

Description of the mechanical properties may be defined as a combination of the relationships governing two-phase composites (gas–solid or reinforcement–solid) with a factor to correct for the volume fraction of each constituent.

### 10.5.1. Determination of Void Content

To define the necessary relationships, some general equations for reinforced foams must first be stated and understood.

Consider a reinforced foam of total volume $V_c$, which consists of three components, $V_m$ the matrix volume, $V_r$ the reinforcement volume and $V_g$ the gas volume. If the volume fraction, $\phi$, of each component is similarly denoted, the following relationships can be defined:-

$$V_r + V_m + V_g = V_c \tag{10.1}$$

$$\phi_r = \frac{V_r}{V_c}; \quad \phi_m = \frac{V_m}{V_c}; \quad \phi_g = \frac{V_g}{V_c} \tag{10.2}$$

Let us now define $\phi_g^*$, the gas content of a gas–solid system:

$$\phi_g^* = \frac{V_g}{V_g + V_m} \tag{10.3}$$

The introduction of two further concepts is now required. The density ratio, $\beta$, and the mass ratio, $n$, may be defined:-

$$\beta = \frac{\rho_m}{\rho_r} \tag{10.4}$$

and

$$n = \frac{\text{mass of reinforcement or filler}}{\text{mass of matrix}} = \frac{\phi_r \rho_r}{\phi_m \rho_m} \tag{10.5}$$

The masses of the phases are denoted by $m_r$, $m_m$ and $m_g$ and the mass of the composite by $m_c$.

By appropriate rearrangement of eqns (10.1)–(10.5), relationships for $\phi_v$, $\phi_r$ and $\phi_g^*$ in terms of $\beta$, $n$ and the densities of the constituent phases may be obtained. These equations (10.6), (10.7) and (10.8), are derived below.

(i)
$$\phi_g = 1 - \phi_r + \phi_m$$

$$= 1 - \frac{\rho_c(V_r + V_m)}{m_c}$$

$$= 1 - \frac{\rho_c(V_i/V_m + V_m/V_m)}{m_c/V_m}$$

Assuming that $m_c = m_r + m_m$, i.e. $m_g = 0$, which is obviously sensible:

$$\phi_g = 1 - \frac{\rho_c(\beta n + 1)}{\rho_m(n + 1)} \qquad (10.6)$$

(ii)
$$\phi_r = \frac{V_r}{V_c}$$

$$= \frac{\rho_n m_m}{V_c \rho_r} \quad \text{since} \quad V_r = \frac{n m_m}{\rho_r}$$

It may be shown that $\dfrac{m_m}{V_c} = \dfrac{\rho_c}{(n + 1)}$ assuming once again that $m_g = 0$.

Thus
$$\phi_r = \frac{\rho_c}{\rho_r(n + 1)} \qquad (10.7)$$

(iii)
$$\phi_g^* = \frac{V_g}{V_g + V_m} = \frac{V_g}{V_c}$$

In this case the gas bubbles may be considered as the filler and so:

$$n = \frac{\text{mass of gas}}{\text{mass of matrix}}$$

Since
$$n = \frac{\rho_g V_g}{m_m}$$

$$\phi_g^* = \frac{n m_m}{\rho_g V_c}$$

$$\frac{m_m}{V_c} = \frac{\rho_c}{n + 1}$$

Thus
$$\phi_g^* = \frac{\rho_c}{\rho_g(n + 1)} \qquad (10.8)$$

These relationships are summarised in Fig. 10.4.

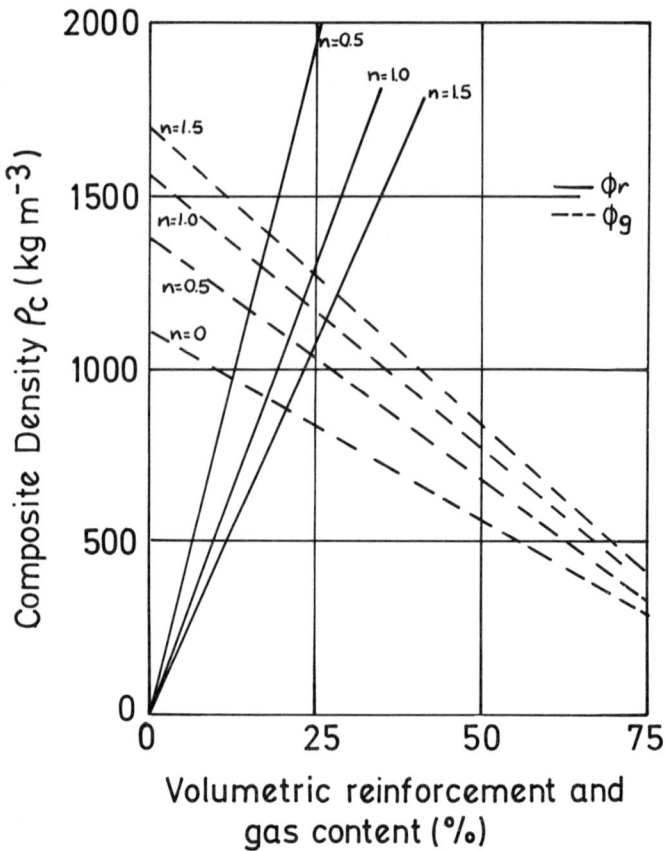

**Fig. 10.4.** Evaluation of three-phase composite parameters $\phi_r$ and $\phi_g$ from composite density and weight ratio, $n$; $\rho_r = 2.6 \times 10^3$ kg m$^{-3}$, $\rho_m = 1.1$ kg m$^{-3}$.

## 10.5.2. Determination of Modulus

It has been demonstrated[6] that only in the case where all components have the same bulk modulus does the simple rule of mixtures hold. Kerner[6] and Halpin and co-workers[7] derived a series of equations to describe unidirectional, discontinuous fibre reinforced composites and these show that:

$$\frac{E_{c11}}{E_m} = \frac{1 + A\eta\phi}{1 - \eta\phi} \tag{10.9}$$

where  $E_{c11}$  = Young's modulus of the composite;
  $E_m$  = Young's modulus of the matrix;
  $A$  = twice the aspect ratio of the reinforcement, and

$$\eta = \frac{(E_r/E_m) - 1}{(E_r/E_m) + A} \tag{10.10}$$

The transverse modulus, $E_{c22}$, and the modulus of spherical reinforcements can be expressed by eqn (10.9) by letting $A = 2$.

The derivations need to be expressed for a randomly reinforced composite (as that is what a reinforced thermoplastic foam effectively is) and a linear combination of the equations for $E_{c11}$ and $E_{c22}$ gives:[8]

$$E_c = \tfrac{3}{8} E_{c11} + \tfrac{5}{8} E_{c22} \tag{10.11}$$

Considering this via $E_{c22}$, as stated above, with $E_r = 0$ and $A = 2$,

$$\frac{E_f}{E_m} = \frac{1 - \phi_g^*}{1 + \phi_g^*/2} \tag{10.12}$$

where  $E_f$ = Young's modulus of the thermoplastic foam and $\phi_g^*$ = volume fraction of gas in the two-phase composite.

$$\frac{E_{c11}}{E_m} = \frac{1 + A\eta\phi_g^*}{1 - \eta\phi_g^*} \tag{10.13}$$

$$\frac{E_{rf}}{E_{c11}} = \frac{1 - \phi_g}{1 + \phi_g/2} \tag{10.14}$$

where $E_{rf}$ is the modulus of the reinforced thermoplastic foam. Combining eqns (10.12) to (10.14):

$$\frac{E_{rf}}{E_f} = \frac{1 + \eta A\phi_g^*}{1 - \eta\phi_g^*} \times \frac{\dfrac{1 - \phi_g}{1 + \phi_g/2}}{\dfrac{1 - \phi_g^*}{1 + \phi_g^*/2}} \tag{10.15}$$

If the unreinforced foam has the same density as the matrix component of the reinforced foam, i.e. $\phi_g = \phi_g^*$, then the second term on the right-hand size of eqn (10.15) disappears and the relationship is analogous to the original Halpin–Tsai equation. The volume fraction of reinforcement is not taken directly from the composition of reinforced foam but is given by eqn (10.8).

Although $E_{rf}$ gives the modulus of the reinforced foam in the injected direction, the analysis does not account for the more complex equivalent transverse modulus.

A simplified approach to the model is given by application of the rule of mixtures.[9]

$$E_{rf} = k \left[ \left( 1 - \frac{\% \text{ Foam}}{100} \right) E_{cll} + \left( \frac{\% \text{ Foam}}{100} \right) E_a \right] \qquad (10.16)$$

where   $k$ = efficiency constant and
    $E_a$ = Young's modulus of air, which is negligible with respect to the modulus of the reinforced thermoplastic matrix.

Thus

$$E_{rf} = k \left( 1 - \frac{\% \text{ Foam}}{100} \right) E_{cll} \qquad (10.17)$$

Therefore, as the foam level increases, the modulus of the glass fibre reinforced thermoplastic decreases, and the same assumption can be made for other mechanical properties.

The constant $k$ determines the rate of change of these properties; if $k$ is less than unity, the rate of change of properties will be less than the change in foam level, giving an increased strength-to-weight ratio in the reinforced thermoplastic foam composite, and this is indeed true for glass reinforced thermoplastic foam composites. In fact $k$ is a function of cell size and cell distribution and as such is not a true constant, but it may be considered as such for a given thermoplastic over a small range of foam levels.


## 10.6. TOUGHENING EFFECTS

As discussed in Chapter 2, the presence of a proportion of reinforcing fibres with a length equal to or greater than the critical length significantly improves the toughness of a thermoplastic foam. The economic consideration of this phenomenon results in the use of cheaper fibres, such as glass, to promote toughening while using stronger, more expensive fibres, such as carbon, to promote strengthening. The use of talc as a filler increases stiffness in polypropylene foams.[3]

Reinforced thermoplastic foams absorb impact energy via the cellular structure and, because the cells are discrete energy absorbers, very little crack propagation is possible. This results in a very low fall-off in impact properties over a large range of density reductions.[9] Providing suitable coupling agents are used with the reinforcements, the glass fibres may also act as crack arresters by means of crack blunting mechanisms.

## 10.7. ADVANTAGES OF FIBRE REINFORCEMENT

The addition of glass fibres to a thermoplastic foam increases the apparent viscosity and the thermal conductivity resulting in improved cell size control, shorter moulding cycles and hence more uniform mechanical and physical properties.

Unreinforced thermoplastic foams are suitable only to replace metals and other materials in low-stressed applications,[2] but the addition of glass fibre reinforcements to the foam produces a composite ideal for more highly stressed conditions. The special features of these glass fibre reinforced thermoplastic foams are numerous. They include the following.

(i) Glass fibres increase the strength of the cellular structure, adding back the strength normally lost when unreinforced thermoplastics are foamed. The glass fibres orientate themselves at random tangentially to each cell wall (Fig. 10.5), thus reinforcing it individually.

(ii) They are relatively free from moulding stresses. In production the mould is filled by internal expansion of the material,

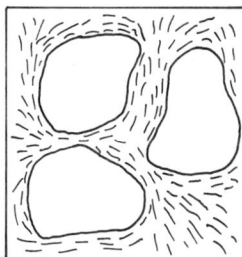

Fig. 10.5. Glass fibre reinforced thermoplastic foams showing reinforcement distribution in relation to the cells.

minimising orientation of the thermoplastic molecular structure and hence orientation and packing stresses in the component.

(iii) They are ready to mould on existing machines. Glass fibre reinforced thermoplastic foams can be moulded on existing injection moulding machines with a positive shut-off nozzle to prevent the escape of gas from the cylinder.[10] Chemical blowing agents compounded within the resin are activated in the moulding process and expand to form the cellular structure within the component.

(iv) They have high strength-to-weight ratio and excellent creep resistance. When reinforced thermoplastics are foamed, properties decrease proportionally to the amount of foaming, but remain several times superior to those exhibited by the analogous thermoplastic foam without reinforcement. The tensile strength of the foam may be increased by a factor of 1·6–2·5 by the addition of glass fibres,[2] depending on the particular thermoplastic being reinforced.

(v) Fast cycle times are possible. Glass fibre reinforcements stiffen the outer skin of the foam, a result of the higher modulus of glass fibre reinforced thermoplastics, which allows earlier ejection of the components which can then be cooled out of the mould. Post-blowing or cauliflowering of the component must however be avoided if this method is used.

(vi) Retention of nails and screws is improved. As the nail or screw enters the cellular structure of the component, cells collapse around it, and this creates better retention characteristics. Furthermore, as glass fibre reinforced thermoplastics tend to peen when impacted and therefore do not propagate cracks so readily as their unreinforced counterparts, they may be nailed together when it would be disastrous to nail conventional thermoplastics.

Components can be designed with fewer ribs and bosses because of the improved nail and screw retention characteristics. This also helps reduce cycle time.

(vii) Warp-free components are easily produced. When moulding with high pressures, as in conventional injection moulding, internal stresses are introduced into the components, and this frequently produces warpage after ejection. However, the low pressures used to mould reinforced thermoplastic foams produce very little stress in the components and warp is almost eliminated.

(viii) Sink marks are virtually eliminated. The material is allowed to expand freely to fill the mould completely, and the resultant internal pressure in the melt produces a sink-free component.

(ix) Thermal insulation is increased Frequently the insulation properties of the reinforced thermoplastic foam are superior to those of the analogous solid moulding, but the significant reductions in thermal conductivity necessary for true insulation applications are only possible at great density reductions and this results in significant losses in physical and mechanical properties.

(x) Acoustic damping is increased. Reinforced structural foam components provide useful acoustic damping which has obvious benefits. Vibration noise can also be minimised.

(xi) Lower tooling costs are possible. These may be achieved by the use of cast aluminium as a mould material, this being possible because of the low internal pressures necessary in the mould. Particular advantage is obtained in short-run applications, where high tooling costs would otherwise preclude the use of injection moulded reinforced thermoplastic foam. Glass fibres are however very abrasive and this may offset any advantage gained from lower tooling costs if continuous or semi-continuous production is required because of the down time in mould changes. The savings do however depend on the thermoplastic being processed and the tolerances required in the final component.

(xii) Reinforced thermoplastic foams produced by low-pressure systems tend to have a characteristic swirled surface finish, which is desirable, when suitably pigmented, for wood grain effects. A textured mould finish also helps to complement this finish, but care must be taken not to introduce notches, which would act as stress raisers, reducing the impact properties of the component.

## 10.8. FINISHING PROCESSES

It is vital to allow reasonable time between demoulding and finishing so that any residual pressure in the cellular structure equilibriates and does not disturb the subsequent operations. This time is dependent upon the reinforced thermoplastic being produced.

Painting is preferred for finishing, but the swirled surface finish may be a hindrance to this, if it is necessary to produce a smooth surface finish. A smoother finish is produced by raising the mould temperature. More expensive and sophisticated machines are available which enable further enhancement of surface finish. If a high quality paint finish is required, then surface preparation, priming, high built base coats and even sanding before the final finishing coat are necessary. This is expensive and the economics of these operations often mean that it is cheaper to improve the surface finish by improved moulding techniques as discussed in Section 10.2.2.

## 10.9. THERMOPLASTICS VERSUS THERMOSETS

The processing of both reinforced thermoplastic and thermoset foams is similar, with relatively low mould pressures being involved in both systems, the working pressure arising from the degassing of the melt or liquid and not from the injection pressure.

The most obvious difference between the two systems is that thermoplastics are used in pellet or powder form whereas the thermosets are frequently liquid systems.

Most versatile and widely used of the thermosets are polyurethanes. The physical and mechanical properties can be tailored by suitable density reductions and by adding reinforcements to produce components suitable for a wide range of applications.

However, a comparison of the two types is complex and generalisations are misleading and unrealistic, so that each application must be considered and evaluated individually, with particular reference to:

    (i)   required surface quality;
    (ii)  required strength;
    (iii)  temperature resistance; and
    (iv)  flame resistance.

Thermoplastics have the advantage that, under normal operating conditions, any rejected components can be recycled by grinding and further processing with new polymer, particularly if the subsequent component surface is to be painted or foil coated.

## 10.10. APPLICATIONS

The excellent mechanical and physical properties, together with the dimensional stability, of reinforced thermoplastic foams provide a wide variety of applications in fields which were originally thought to be the domain of unreinforced foams[3] until inherent failings were fully realised, e.g. considerable losses in mechanical properties.

The addition of fibre reinforcements has offset many of these failings, and designers, moulders and material suppliers tailor the additives to achieve the most suitable composite for a given application. Common additives are glass fibres, talc, calcium carbonate and rubber modifiers. Not only are the mechanical and physical properties improved, but the additives also act as nucleation sites, promoting cell formation and uniformity during moulding. Short glass fibres are at present the main source of reinforcement in structural thermoplastic foams, although carbon and synthetic fibres are also used.

With these factors in mind, the feasibility of using reinforced thermoplastic foam for moulding the cylindrical outer tank of a front-loading washing machine was considered by ICI, Philips and Cabinet Industries.[11, 12] The inherent design difficulties involved in the manufacture of a horizontally rotating drum include large investments in presses, tools and welding equipment when the drum is produced in metal. The two main alternatives are vitreous enamelled mild steel or stainless steel that is reinforced by steel components to carry the load. To replace the separate operations, 16 in all, by one single moulding is obviously very attractive, lowering the capital investment considerably, but to do this the designers had to ensure that the tank was capable of fulfilling the necessary requirements of watertightness, corrosion resistance and high strength to carry the washing load and ancillary equipment.

The material selected was fibre glass reinforced polypropylene, which has outstanding chemical resistance, good heat stability and resistance to abrasion and fatigue. An added advantage is the reduction in noise level from the machine because of the acoustic deadness of polypropylene. This design has proved very successful and other washing machine manufacturers are now involved in development of reinforced thermoplastic drums, although not all developments are concerned with thermoplastic foam but with solid resin components. This is due to the possibility of achieving faster cycle times with certain designs of solid mouldings.

Figure 10.6 compares the old vitreous enamel drum with a new reinforced polypropylene foam drum. Although this represents a large part of the present reinforced thermoplastic foam market, other applications are numerous and may be classified into four areas of usage.

**Fig. 10.6.** The original vitreous enamelled tank (*left*) compared with the reinforced polypropylene foam version (*right*), showing the complexity of moulding in the drum. (Courtesy Imperial Chemical Industries PLC.)

### 10.10.1. Static Loading

Successful applications are the desk support moulded by Hollis Plastic Ltd giving a functional component with good rigidity and stiffness at reasonable cost, and the Sharna Ware wheelbarrow. Although obviously mobile, the main loading is static and the wheelbarrow has a rigid framework with high stiffness. It is tough, resilient and light. The wheelbarrow is offered in a choice of colours, which remain attractive after prolonged outdoor use, all features difficult to achieve with metals. Chairs produced by Wetherell Plastics are intended for the large-quantity prestige market and need to be rigid with good stiffness while being cost effective and available in various styles.

### 10.10.2. Dynamic Loading

Probably the first commercial application of glass reinforced thermoplastics foam was the 'ACE' canoe paddle, designed by Robin Witter, the British double canoe slalom and Olympic coach. The major design features were low weight, good tensile and fatigue strengths, flexural stiffness, impact resistance to sharp objects and weather resistance.

### 10.10.3. High Temperature

The Ecko Hostess 'Carousel' is an electrically heated food server and warmer, where the heating element is attached to an aluminium dish and mounted inside the moulded base. The moulding must be capable of withstanding an operating temperature of 85°C and all fats and cleansing agents, while supporting all the other components. Glass fibre reinforced polypropylene foam which can also be attractively finished, is used.

The electrical industry also uses electric brake controllers, with resistors that can reach 200°C, manufactured from suitable glass fibre reinforced thermoplastic foams.

### 10.10.4. Combined Conditions

Designers in the automobile industry have successfully utilised reinforced thermoplastic foam in several areas:[3]

(i)   glass fibre reinforced polycarbonate foam to manufacture door sills and certain roof sections;
(ii)  instrument panels, permitting reduced panel complexity;
(iii) seating for trucks and tractors;
(iv)  bonnets and exterior panels.

The trend towards reduction of interior noise levels in commercial vehicles favours the use of housings and panels manufactured from polypropylene foam because the stiffness and mechanical inertness of the panels effectively damps out the transmitted vibrational noise to which metal panels are particularly prone. It is not, however, normally necessary to reinforce these mouldings.

### 10.10.5. Summary of Applications

Practically any foamable thermoplastic can be reinforced by up to 30% fibres. Suitable thermoplastics include high-density polyethylene (HDPE), polypropylene (PP), polystyrene (PS), acrylonitrile–butadiene–styrene (ABS), styrene–acrylonitrile (SAN), polycarbonate (PC), polyphenylene oxide (PPO) and nylon. Table 10.2 summarises the main applications.

TABLE 10.2
General Applications of Fibre Reinforced Thermoplastic Foams

| Application | Glass fibre reinforced foam | | | | | | | |
|---|---|---|---|---|---|---|---|---|
| | Nylon | PS | ABS | SAN | PP | HDPE | PPO | PC |
| Furniture and panels | | √ | √ | √ | √ | √ | | |
| Building products | | | √ | √ | √ | | | |
| Domestic appliances | √ | √ | √ | √ | √ | | | √ |
| Electrical components | √ | | √ | | | | | √ |
| Sports equipment | | | | √ | √ | | | |
| Housings and components for business office machines | | | | | | | √ | |
| Automobile seats and components | √ | | | | √ | | | |
| Material handling/ crates | | | | | √ | √ | | |
| Telephone distribution boxes | | | | | | | | √ |

## 10.11. FUTURE DEVELOPMENTS

Future growth in the applications of reinforced thermoplastic foams depends primarily on the automobile and domestic appliance markets and these industries have a sufficient flow of new models to justify trials with new engineering materials. A significant move is the use of solid mouldings wherever feasible in order to reduce moulding times, particularly in automobile seating.

Generally the future of reinforced thermoplastic foams is uncertain. However, there is no doubt that where low densities combined with good mechanical properties are required, they are very strong candidate materials.

## REFERENCES

1. Santrach, D., *Polymer Composites,* 1982, 3, 239–44.
2. Mandy, F., *Plastics and Rubber,* 1976, 1, 119–25.

3. Dominick, G., *J. Elastomers and Plastics,* 1979, **11**, 133–9.
4. Throne, J. L., 'Effect of cellular structure and chemical foaming agents on resin properties in the almost-solid region', *SPE Annual Technical Conference, New Orleans,* 1979.
5. Throne, J. L., *Mechanics of Cellular Polymers, Structural Foams,* Applied Science Publishers, London, 1981.
6. Kerner, E. H., *Proc. Phys. Soc.,* 1956, **69B**, 808–16.
7. Halpin, J. C. and Kardos, J. L., *Polym. Eng. Sci.,* 1976, **16**, 344–52.
8. Ashton, J. E., Halpin, J. C. and Petit, P. H., *Primer on Composite Materials,* Technomoc, Westpoint, Conn., 1969.
9. Wilson, M. G., 'Glass reinforced thermoplastic foam', *SPE Regional Conference, South California,* March 1972.
10. Mount, R. K., *Fibreglass Reinforced Thermoplastic Structural Foam,* SAE paper 720478, May 1972.
11. Waterman, N. A. and Pye, A. M., *Mater. Eng. Appl.,* 1979, **1**, 203–208.
12. Norgan, M., *New Washing Machine Uses Propathene Structural Foam,* ICI PLC, Welwyn Garden City, 1979.

# Index